THE PROCESS ENNEAGRAM©

Essays on Theory and Practice

THE PROCESS ENNEAGRAM©

Essays on Theory and Practice

Edited by Richard N. Knowles

EMERGENT ™
PUBLICATIONS
3810 N 188th Ave
Litchfield Park, AZ 85340

The Process Enneagram©: Essays on Theory and Practice

Edited by: Richard N. Knowles

Library of Congress Control Number: 2013945955

ISBN: 978-1-938158-10-0

Copyright © 2013 3810 N 188th Ave, Litchfield Park, AZ 85340, USA

Printed in the United States of America

ABOUT THE AUTHORS

Richard Bergeon is a principal of Bergeon, Fu and Associates. He holds a BBS in General Management (Wayne State University), an MA in Whole Systems Design (Antioch University), and certification as an Organization Systems Renewal Consultant. He has been employed as manager, director, and executive leadership consultant in banking, public utilities, transportation, manufacturing, pharmaceutics, and telecommunications, specializing in the adoption of new technologies and staff development. He is presently pursuing his doctoral degree at Gonzaga University, focusing on the role of values in global and intercultural leadership.

Anthony Blake was born England 1939. B.Sc. Physics Bristol University, Certificate in History and Philosophy of Science, Cambridge. Collaborator with creative engineering designer Edward Matchett. Management consultant, teacher and editor/publisher. Development of systematics, structural communication and logovisual technology. Cofounder of DuVersity. Member of ANPA (Alternative Natural Philosophy Association, devoted to implications of combinatorial hierarchy). Collaborator with group analysts de Mare and Lawrence and practitioner of Bohmian dialogue. Published books include: *A Seminar on Time, The Intelligent Enneagram, The Supreme Art of Dialogue, A Gymnasium of Beliefs in Higher Intelligence.*

Caroline Fu is assistant professor of the Doctoral Program in Leadership Studies at Gonzaga University. She holds a PhD in Leadership and Change and an MA in Whole System Design, both from Antioch University. Her MS in Computer Sciences and BS in Applied Mathematics, Engineering (EE) and Physics are both from University of Wisconsin. She has a Certificate of Completion in System Dynamics Advanced Study from the Sloan School of Management, Massachusetts Institute of Technology. Her passion is furthering Leadership-as-Energy-Flow concepts for assessing complexity from a Tao leadership and modern physics lens for leadership decision support and global policy learning.

Beverly McCarter is an award winning architect/designer of 3D Immersive Virtual Spaces. Her work focuses on the psychology of the avatar and virtual worlds, "wicked problems", the complexity of immersive learning spaces, as well as the impact of the aesthetics of 3D immersive environments on complex human systems. Experience includes US Army Simulation and Training Technology Center (Virtual Environment Consultant), the National Defense University (Federal Consortium for Virtual Worlds Program Manager, Education Innovation Coordinator), the Smithsonian Institute (Consultant; Executive Coach, Organizational Analyst, Group Facilitator) and Innovative Decisions, Inc. (Virtual Environment and Complex Systems Consultant).

Mark McGibbon is a visiting graduate school professor at the National Defense University (NDU) Information College. He works with executive Strategic Planners at Lockheed Martin. Prior to returning to the U.S. in 2006 after living years in European countries, he worked as the Lockheed Martin European Operations Officer and European Chief Enterprise Architect. Dr. McGibbon worked for the U.S. government and corporations as a Chief Information Officer, Project Mgr, Program Mgr, Regional Mgr, Operations Officer, Strategic Planner and founded small businesses. He has taught undergraduate, graduate and doctoral business and technology courses as an adjunct professor since 1996 at the U of West Florida, Troy State U, U of Maryland U College (London, UK & Stuttgart, Germany), CSU, Pensacola Junior College, and Northcentral University. He is a member of Delta Mu Delta Business Honor Society and Distinguished Graduate recipient after earning 15 Graduate School Certificates in Strategic Leadership, Organizational Transformation, Chief Information Officer, Chief Technology Officer, Chief Financial Officer, Program Mgt, Project Mgt, Chief Information Security Officer, Enterprise Architect, Chief Enterprise Architect, Teaching and multiple Information Assurance certificates. He holds a Ph.D. in Business Administration, Doctorate in Business Administration, M.S. in IT Mgt, M.S. in Strategic Leadership, and a B.S. in Political Science (pre-law). He is a Harvard University Senior Executive Fellow.

Dr. McGibbon currently teaches in the following NDU courses and programs: Advanced Management Program (AMP); Analytics and Simulation for Enterprise Architects (ASA); White House, Congress, and the Budget (BCP);Enterprise Architecture Capstone (CAP); Changing World of the CFO (CFF); Continuity of Operations (COO); Cornerstone Seminar (CRN) ; Critical Information Systems Technologies (CST); CTO Roles and Responsibilities (CTO);Cyberspace Strategies (CYS); Domestic Field Studies (DFS); Emerging Technologies (EIT); Local Field Studies (LFS); Global Enterprise Networking and Telecommunications (GEN); National Intelligence and Cyber Policy (NIC), Organizational Culture for Strategic Leaders (OCL); Risk Management, Internal Controls and Auditing for Leaders (RIA); Strategies for Assuring Supply Chain Security (SAC), Software Acquisition Leadership (SAL), and Web-Enabled Government: Facilitating, Collaborating, and Transparency (WGV).

Dr. **Cameron Richards** is an Australian academic with extensive experience of working in the Asia-Pacific region—including positions at Nanyang University Singapore, the Hong Kong Institute of Education and the University of Western Australia. He has a multi-disciplinary background which includes specializations in academic research and writing methodology, educational technologies, intercultural communication, curriculum innovation, and new literacies. In his 15 years or so of focusing on new approaches to higher as well as school education he has developed a particular interest in the development of sustainable policy-building research and strategies in wider social as well as organizational contexts—now with a particular focus on 'science, technology and innovation' policy studies and research inquiry.

Paul Rowland has held various levels of management positions for large organizations in the telecommunications industry. His introduction to management came in 1985 when he was tasked with a project to provide a wide area data network for a new customer service computer system. The first day on the job taught him that the strongest resistance to change can come from within the management of an organization. His second early lesson was managers always want more; more people, more resources etc. From this experience came a desire to get more out of existing resources to be able to develop a high performing team. In 1988 he left England with his family to live and work in Canada. The first two years were spent in Ontario managing a team of field technicians spread over a wide geographical area. From this experience came the challenge of creating a sense of common purpose, and a sense of belonging within a group who had little daily contact with one another but even less contact with the broader organization. In 1990 he moved to British Columbia to take the position of Director. During his time as Director, Paul experimented with various management concepts and ideas. The unfortunate guinea pigs were his management team. Their minds were probed with various personality instruments, they experienced 360 feedback surveys, they ran an airline for a day to learn about systems thinking, they mind mapped and wore "six hats". Some progress was made in turning the team into a more cohesive and productive unit. His experience with Edward DeBono's thinking skills brought the best results in terms of productivity. In 2005 Paul turned his attention back to systems thinking, looking for a way to make the team more cohesive. Paul and his team embarked on a series of workshops using the Process Enneagram©, a system tool developed by Richard N Knowles. The team emerged much stronger and with a working framework for continued development. Paul retired in 2008.

Every day **Steffan Soule** expertly shows people that anything is possible! He weaves a visual tapestry of dazzling magic into a variety of theatrical shows: the audience laughs and their heads shake in astonishment as they see lights fly around Steffan and turn into fire. A volunteer narrowly escapes an impossible wrapping of ropes tied through his own jacket and sees a signed $100 bill vanish from his hand, only to appear embedded in a lemon moments later. Steffan's magic is recognized with a Kennedy Center Award for the Arts. He has performed on National Television, twelve times for Bill Gates and in hundreds of corporations nationwide. His show entitled "The Magic of Recycling" is sponsored by the Department of Ecology. Steffan presents the Process Magic Experience to demonstrate the enneagram and the Nine Term Symbol principles from his book Accomplish The Impossible. He gives corporate audiences a practical stimulating approach to process transformation. As designer and co-producer of two million-dollar-magic-theatres custom built for his shows Steffan Soule performed the longest running magic show on the West Coast, "Mysterian". The show's five year run featured some of the greatest magic in the world according to magicians, critics and magic historians. Steffan and his wife, Barbara are based in Seattle where their performances are well known in the arts scene. They have worked with the Seattle Symphony, Pacific Northwest Ballet, Seattle Reparatory Theatre, The Fifth Avenue Theatre, Village Theatre, and Seattle Children's Theatre. Currently, Steffan Soule performs for theatres, corporate events and private parties while creating new works with artist Cooper Edens and their magic ensemble.

Barry Stevenson consults in the fields of Organizational Development (OD) and Strategic Project Planning. His clients include a wide variety of government, private sector, health care and educational institutions. His expertise in systematics and complex adaptive systems is applied in the areas of organizational development and design, coaching and mentoring leaders and facilitating team development. As President, B.W. Stevenson & Associates, Ltd., Barry brings over 30 years of executive leadership and management experience in the health care and high technology sectors with over 10 years as a senior provincial government official in the Provinces of Alberta and British Columbia. As an Associate Faculty member, Royal Roads University and Adjunct Professor, University of British Columbia, he has supervised students in the areas of leadership, systems and complexity theory and organizational change. He has served as an external faculty consultant with the Banff Centre for Leadership and is a Fellow of the Institute for the Study of Complexity and Emergence (ISCE) and an accredited associate of the Center for Self-Organizing Leadership. Barry has a B.Sc. from Simon Fraser University and a M.Sc. (Biochemistry) and M.Sc. (Health Services Planning) from the University of British Columbia. He is an accredited facilitator with qualifications in both HBDI and the application of Systematics. Barry has written numerous scientific/research and management related articles on a variety of topics including program evaluation, management career development, inter-agency/inter-governmental planning, regional health planning, shared-services and information systems partnering, complexity and organizational wellness and complexity and community.

Catherine Taylor started her career as a veterinarian. Realizing the real problems were more to do with the owners and their interaction with the "vets", she re-schooled in the fields of human communication and organizational change. She is recognized as an extremely effective teacher of the micro skills and critical elements of precision communication. She works as a management consultant and trainer with global corporations in Australia, USA, New Zealand and China, in industries as diverse as supply chain logistics, mining, health care and heavy manufacturing.

Hua Wang holds a Ph.D. in Communication from University of Southern California. She is currently an Assistant Professor at the Department of Communication, University at Buffalo, The State University of New York. Her research interests include social structures, communication technologies, and social change. She has presented and published on topics related to social networks, new media, health promotion, and social transformation. She became interested in complexity science in 2010 and has been experimenting with innovative strategies and methods in her teaching and research.

CONTENTS

EDITORIAL: THE PROCESS ENNEAGRAM©
ESSAYS ON THEORY AND PRACTICE

Richard N. Knowles

Introduction 1
References 6

CHAPTER 1
THE TRIPLE ENNEAGRAM

Anthony Blake

Background 9
Structured Process 10
The Primary Relationship 11
   Cybernetics and Learning 13
The Three and the Six 15
REFERENCES 18

CHAPTER 2
OLD WISDOM FOR A NEW WORLD IN CRISIS? THE
ENNEAGRAMMATIC STRUCTURE OF INTEGRATED,
OPTIMAL AND SUSTAINABLE PROBLEM-SOLVING

Cameron Richards

Introduction 22
The Unity of One (or More) as Symbol for the Inherent Universality
   of Specific Systems 24
The Relation Between Complex Problems and the Enneagram's
   Depiction of the 'Law of Three' 29
The Enneagrammatic Formula for Integrated, Optimal, and
   Sustainable Problem-Solving 34
Conclusion 39
References 40

CHAPTER 3
2008 MECS SUMMIT: A WORKSHOP ON COMPLEX SITUATIONS

Beverly G. McCarter

MECS Forum 44
Structure of the Workshop 45
    Opening Remarks 46
Using the Process Enneagram© for MECS 49
The Process Enneagram© Maps for the MECS Workshop ... 52
Key Learning from this Use of a Modified Process Enneagram© 63
Acknowledgements 65

CHAPTER 4
THE PROCESS ENNEAGRAM©
A PRACTITIONER'S GUIDE TO ITS USE AS A FACILITATIVE TOOL IN THE CORPORATE ENVIRONMENT

Catherine Taylor

Introduction 67
Caution 69
Strengths 70
A whole-of-system perspective 70
Application 71
Where it doesn't work so well and what to do about it 76
Who should use it 78
Authenticity and duration 78
Diagnosis 80
Reprise 81
References 82

CHAPTER 5
HOLISTICALLY EDUCATING GRADUATE STUDENTS FOR THE CONCEPTUAL AGE USING THE PROCESS ENNEAGRAM©

H. Mark McGibbon

References 90

CHAPTER 6
HOLISTICALLY EDUCATING GRADUATE STUDENTS USING THE PROCESS ENNEAGRAM©

Hua Wang

Introduction..94
Complexity Science and the Process Enneagram 94
Applying the Process Enneagram in a University Classroom..............96
    Introducing The Process Enneagram..96
    Running The Process Enneagram 97
    Initial Feedback On The Use Of The Process Enneagram..............99
Revisiting the Process Enneagram 101
Course Evaluation on the Use of the Process Enneagram 102
Lessons Learned and Further Implications.....................................103
Acknowledgement 104
References..104

CHAPTER 7
TOOLS OF COMPLEXITY: THE PROCESS ENNEAGRAM©

Barry W. Stevenson & Paul Rowland

In the Beginning...108
Outcomes Associated with Team Development Work.....................114
    The Influence of Structure/Context on Effective Adoption of
        the Team Goals, Principles, Relationships and Information Sharing
        Components of the Process Enneagram..................................114
    The Didactic Nature and Process of Getting to an Agreed
        Set of Principles/Values...116
    The Rationalization of Intention and Work in Achieving Effective
        Work Performance and its Effect on Other Components.......119
    Dealing with Tensions/Issues in an Integral Manner Through the
        Process Enneagram and Seeing the Effect On and/or Contribution
        of Other Components...120
Key Success Factor—Principles and Standards 121
Observations for future work...122
Concluding Remarks 124
References..125

CHAPTER 8
A TAO TRANSFORMATION LEADERSHIP MODEL
FOR THE PROCESS ENNEAGRAM©

Richard Bergeon & Caroline Fu

A Tao Transformation Leadership Model for the Process Enneagram............... 128

    The Tao.. 129

    The Command and Control Paradigm... 133

    The Tao Model and Transformation 135

    Imported Energy Influences and the Process Enneagram 136

Rebalancing and the Tao Model.. 137

Summing up 142

References... 142

CHAPTER 9
PROCESS MAGIC WITH THE ENNEAGRAM

Steffan Soule

Introduction... 146

    Picturing the Scenario... 147

    Steffan Soule's Even and Odd Miracle.. 149

    Pondering the Symbols.. 153

References... 158

THE PROCESS ENNEAGRAM©: ESSAYS ON THEORY AND PRACTICE

Richard N. Knowles

INTRODUCTION

T hese essays are a collection of papers from eleven people who have studied and used the Process Enneagram sharing their various experiences and insights developed over the years. There was such a strong response to the Call for Papers for the Special Issue of *Emergence: Complexity & Organization* (*E:CO*) that this companion book was published including these first four papers as well as papers from five other authors.

The Process Enneagram is a comprehensive tool of complexity enabling people to come together in a structured conversation to transform themselves and their organization. Its use enables people to solve complex problems, build the connections with others with whom they'll need to work and releases the emotional energy and commitment to do the work quickly and well, ALL at the SAME TIME!

The Process Enneagram is an ideal tool for leaders to use to help to transform their organizations. This is the central tool for Self-Organizing Leadership©. WHEN THE PEOPLE COME TOGETHER TO CO-CREATE THEIR FUTURE EVERYTHING CHANGES. Keeping their work posted, discussing together in an ongoing way, sharing the information, modifying it as things change and keeping it alive enables the organization to build a sustainable future. Using the Process Enneagram this way is the most effective, successful, reliable organizational change tool this author knows about.

The recent paper by Axleandros G. Psychogios and Saso Garev in *E:CO* (2012) is a step towards Self-Organizing Leadership. The use of the Process Enneagram will move you much further and faster.

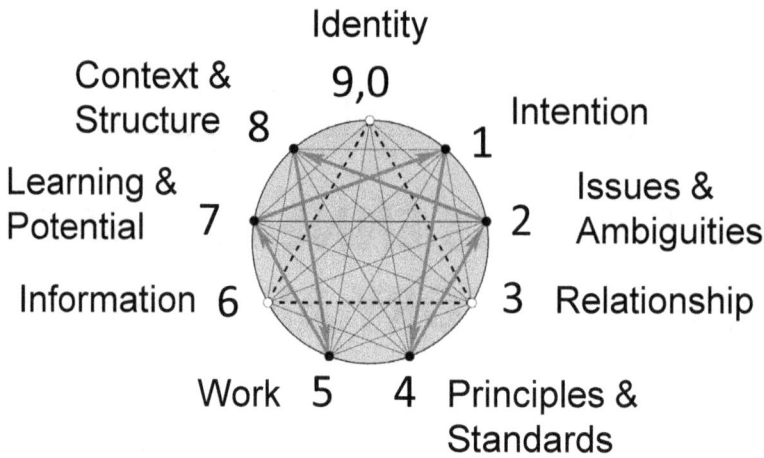

Figure 1 *The Process Enneagram©*

The enneagram was introduced into the West by G. I Gurdjieff in about 1915, and first, publicly written about by his student, Peter D Ouspensky (1949).

Richard N. Knowles was first introduced to the unlabeled and obscure enneagram figure in about 1984 by some people in the DuPont Company who never explained it, but were familiar with the work of John G. Bennett (1983, 1987), a British philosopher and student of Gurdjieff. Knowles spent a number of years reading about the ennea-gram, trying to make sense of it. One key discovery Knowles (2002: 28, 108-111) made was the realization that by placing the terms of Bennett's Systematics around the en-neagram figure, and then renaming them, transformed the figure into tool that is use-ful in studying organizations, in revealing how they work and in opening up the path-way for the people in the organization to transform themselves and their organiza-tion. (The Monad dealing with wholeness was placed at point 1, the Dyad dealing with duality was placed at point 2, the Triad dealing with reconciliation was placed at point 4, the Tetrad dealing with work was placed at point 5, the Pentad dealing with poten-tial was placed at point 7 and the Hexad dealing with structure was placed at point 8.) This figure, with the names placed at each point, Knowles called The Process Ennea-gram©. The term "Process" was used to differentiate it from the more widely known enneagram of personality that is a very different and unrelated use of the enneagram. Another key discovery was the realization that the inner triangle of the Process Ennea-gram figure connecting points 0, 3, 6 and back to 9, was the Self-Organizing Leader-ship Process (Knowles, 2002: 28, 41), and that the Process Enneagram was a powerful tool for use in the arena of complexity that is discussed later in this Introduction.

Here is a brief description of the points:

- Point 0 (Identity): Who are they? What is their Identity? What is their history, individually and collectively? (As the first cycle is completed, this point becomes point 9 as they complete the first cycle to their new Identity).

- Point 1 (Intention): What are they trying to do? What are their Intentions? What is the future potential?

- Point 2 (Issues): What are the problems and issues facing them? What are their dilemmas, ambiguities, paradoxes and questions?

- Point 3 (Relationship): What are their Relationships like? How are they connected to others they need in the system? What is their level of trust and interdependence? What is the quality of these connections? Are there too many or too few of these connections?

- Point 4 (Principles and Standards): What are their Principles and Standards of behavior? What are their ground rules, really? What are the undiscussable behaviors that go on, over and over? What are their espoused values and values-in-use? Is there agreement?

- Point 5 (Work): What is their Work? On what are they physically working?

- Point 6 (Information): Do the people know what's going on? How do they create and handle Information? Who has access to it? Do people understand the information?

- Point 7 (Learning): Are they Learning anything? What are their Learning processes? What new insights have emerged? What is their future potential and new possibilities?

- Point 8 (Structure and Context): How are they organized? What is their Structure? Where does the energy come from that makes things happen in their organization? Is their hierarchy deep or flat? What's happening in the larger environment, in which they're living and trying to thrive? Who are their competitors and what are they doing? What is the Context or surrounding environment in which they are living and working?

- Point 9 (Their New Identity): As they have moved through these questions, how has their Identity changed? Have they expanded and grown? What new things do they now know? What new skills do they now have?

The Process Enneagram works whenever a group is working on a problem that is in the arena of complexity (see Figure 2). In my view, whenever people need to come

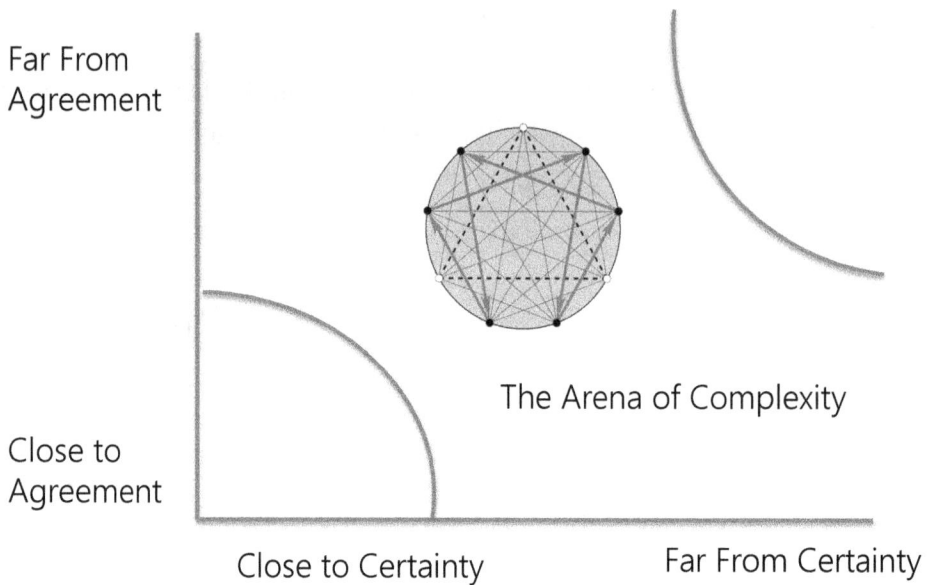

Figure 2 *The Arena of Complexity. Adapted from Stacey (1996).*

together to work on a difficult, knotty problem, they are in the arena of complexity. The people need to come together to be in dialogue around the problem. As they engage this way, using the Process Enneagram, a sense of the whole develops, new ideas emerge, new Principles and Standards of behavior are co-created, and their transformation begins to take place.

The only reason that Knowles relentlessly pursued and developed the Process Enneagram is because it WORKS. In using it, the people working together move back and forth into themselves, and between their organization and the outside world. This back and forth, cyclical process of our personal and organizational transformation enables people to function more effectively in our ever changing, complex world.

Richard N. Knowles first introduced the Process Enneagram, a tool of complexity, to E: CO readers in 2001.

- Richard N. Knowles (2001). "Self-organizing leadership: A way of seeing what is happening in organizations and a pathway to coherence," *Emergence*, 3(4): 112-127.

- Richard N. Knowles (2002). "Self-organizing leadership: A way of seeing what is happening in organizations and a pathway to coherence (Part II)," *Emergence*, 4(2): 86-97.

- Ken Baskin (2002). "A review of Richard N. Knowles, *The Leadership Dance, Pathways to Extraordinary Organizational Effectiveness*," *Emergence*, 4(4): 89-100.

Since these first papers, other papers relating to the use of the Process Enneagram have been written.

- Richard N. Knowles (2006). "Engaging the natural tendency of self-organization. The world business academy," *Transformation*, 20(15): August 10.
- Richard N. Knowles (2006). "The Process Enneagram; A tool from 'The Leadership Dance', parts I and II," *The World Business Academy*, Perspectives, 20(1): August 24.
- Tim Dalmau and Jill Tideman (2011). "The middle ground: Embracing complexity in the real world," *Emergence: Complexity & Organization*, 13(1-2): 71-95.
- Gwen Andrews and Richard N. Knowles (2011). "A practical East-West exploration of leadership and learning," *Emergence: Complexity & Organization*, 13(4): 1-17.
- Bruce Waltuck, "Chaos and complexity, applying concepts from complexity science to quality and organizational development," T*he Human Development and Leadership Division*, American Society for Quality.
- Steffan Soule (2011). *Accomplish the Impossible*, ISBN 9780984240517.

This book, "A Collection of Essays On The Process Enneagram©" contains 9 papers building on and expending the work of these earlier papers.

The first paper by Anthony Blake, "The Triple Enneagram" develops the idea of process having an internal structure revealing a deeper understanding of the figure.

The second paper by Cameron Richards, "Old Wisdom for a New World in Crisis" outlines a framework for addressing all kinds of wicked problems adapting the ancient wisdom of the enneagram as a basis for a theory and practice of integrated, optimal and sustainable problem solving.

The third paper by Beverly McCarter , "2008 MECS Summit: A Workshop on Complex Situations", examines the use of the Process Enneagram to examine the underlying principles of wicked problems.

The fourth paper by Catherine Taylor "The Process Enneagram, A Practitioners Guide to its Use as a Facilitative Tool in the Corporate Environment", describes its use in business along with coaching guidelines.

The fifth paper by Helen Wang "Co-Creating a Meaningful, Shared Learning Space: The Use of the Process Enneagram in a University Classroom" looks at the development of a value system and the co-creation of an open and stimulating social environment in the university classroom.

The sixth paper by Mark McGibbon "Holistically Educating Graduate Students for the Conceptual Age using the Process Enneagram" focuses on a graduate school exercise to educate students to think differently about problem solving and transformation.

The Seventh paper by Richard Bergeon and Caroline Fu, "A Tao Transformation Leadership Model for the Process Enneagram" studies the correlation of the energy flows in the Tao Transformation Leadership Model and the Process Enneagram.

The eighth paper by Barry Stevenson and Paul Rowland, "Learning from the Process of Applying the Process Enneagram" describes a 3 year journey using the Process Enneagram as a tool for team development.

The ninth paper by Steffan Soule, "Process Magic with the Enneagram" instructs the reader to follow a card magic routine that demonstrates how the enneagram can be used to understand, track, observe and improve a whole system.

In using the Process Enneagram we are engaged in working together where the conversations and dialogue are dynamical and ever moving.

This tool always works when people are open to thinking in a different way, are facing a compelling, important issue and have the courage, care, concern and commitment to build on and sustain of their efforts.

The Editor deeply appreciates the contributions of these authors to developing a fuller understanding of this highly effective complexity tool.

REFERENCES

Bennett, J.G. (1983). *Enneagram Studies*, ISBN 9781881408185.

Bennett, J.G. (1997a). *The Dramatic Universe Volume 1: The Foundations of Natural Philosophy*, ISBN 9781881408031.

Bennett, J.G. (1997b). *The Dramatic Universe Volume 2: The Foundations of Moral Philosophy*, ISBN 9781881408048.

Bennett, J.G. (1997c). *The Dramatic Universe Volume 3: Man and His Nature*, ISBN 9781881408055.

Bennett, J.G. (1997d). *The Dramatic Universe Volume 4: History*, ISBN 9781881408062.

Knowles, R.N. (2002). *The Leadership Dance, Pathways to Extraordinary Organizational Effectiveness*, ISBN 9780972120401.

Ouspensky, P.D. (1949). *In Search of the Miraculous: Fragments of an Unknown Teaching*, ISBN 9784871876308 (2011).

Psychgios, A.G. and Garev, S. (2012). "Understanding complexity leadership behavior in SMEs: Lessons from a turbulent business environment," *Emergence: Complexity & Organization*, ISSN 1521-3250, 14(3): 1-22.

Stacey, R.D. (1996). *Strategic Management and Organizational Dynamics*, ISBN 9780273725596, p. 47.

Chapter 1

THE TRIPLE ENNEAGRAM

Anthony Blake

The enneagram is both diagram and evocative image. It is based on the idea of process having internal structure. Structured process is at least threefold. The notion of a completing process involves three 'worlds' and includes values. Cybernetics and the ideas of feedback, double-loop learning, and participant-observers illuminate the meaning of the enneagram. Creative engineering, design and innovation do not work by formula. The inner lines of the enneagram designate 'free intelligence' and working from the future, not the past. The essential role of dialogue.

BACKGROUND

The nine-fold representation of process called the enneagram has its roots in traditional ancient use of number, dating back far before the western master Pythagoras. But it branched and flowered probably in the early twentieth century and looks forward rather than backwards. Though often regarded as some kind of 'mystic' design it is actually a bastion of common sense in the strictest meaning of the word—that is, in dealing the sense or meaning that is common across disciplines and not restricted to any expertise. It uses ancient number-thinking as a means to codify complex and diverse situations in ways that can be seen and grasped as a whole.

It is not a recipe for success, but an aid for what some mid-twentieth century engineers called 'observer-participants'—that is, people like us who strive to accomplish worthwhile things and are involved in what we do rather than being separated and detached as clever manipulators. The conceptual core of the symbol is intelligence itself, in its practical sense of being capable of both vision and drive and also of making use of accident and opposition.

People who want to relate to this symbol are advised to dig deeper into what they know and spread the scope and range of 'informative material' they bring to mind—

drawing on as many different kinds of information as they can—being equally open to physics and mythology as well to any current fashionable trends in management and organization theory. Above all, it would be important to acknowledge values beyond the limits of money and power.

Taken in such a way the enneagram can become a repository for the practical wisdom won through thinking, doing and dialoguing. Just bringing it to mind and reflecting on where one is can bring oneself into insight. What follows is a short exploratory essay on making sense of the way in which the enneagram can help us make sense.

STRUCTURED PROCESS

The enneagram can be read as systems diagram or as a kind of picture; as more or less abstract. It offers a way of realizing the structural similarities between apparently different realities; as, for example, between the arts and engineering, management and psychology, and so on. The common denominator is called 'structured process', which we might start thinking of as being designed towards an end, or realizing a purpose. It is supposed that such a structured process is not just a one-off but repeated, allowing for *learning* to take place. This means that there is a pattern to the process which remains the same, while its content changes according to inputs and circumstances.

There are elements that enter the process from outside while other elements are generated internally. The two together produce an output. Structured process involves intelligence. Intelligence can see and hold a pattern together in the face of changing and uncertain circumstances. It is not like a computer programme or set routine. At almost any point it can 'change its *mind.*'

As already implied, the process represented in the enneagram exhibits an internal structure. Provisionally, we can associate this with an apparatus or organism. Internal structure enables the coexistence (or 'management') of *different* processes operating at different *levels*. The concept of 'level' can be illustrated by considering how, in us, digestion of food and thinking go on at the same time but (largely) independently. However, thinking and moving for example though also somewhat independent of each other can combine or cross over (we have to go to the shop to get our bacon).

The concept of levels is important, not least because we want to address the question of 'added-value'; that we are aiming in our structured process to achieve

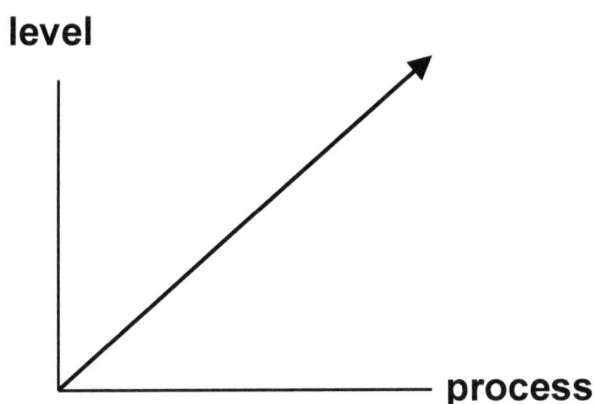

Figure 1 *Simple Picture of 'Worthwhile' Process*

something of *value*, something that will not happen just by itself or by pushing a button. The most primitive picture of this might be:

The qualitative idea of added value is important. It suggests that we start from something given and then improve on it. So there is the idea of a basic—perhaps mechanical or routine—process that is 'added to' by some other process. An obvious illustration is that of a radio frequency which can be modified to carry a signal. This example also leads us to think of how the signal is to be received and made use of, which is then a third level of process.

In the enneagram, process is triple and divided into three phases.

THE PRIMARY RELATIONSHIP

I t's a valuable rule of thumb to look at any worthwhile process in terms of the interlocking of three processes. Take building a house: one starts with making plans and doing calculations; one also has to bring materials and workmen onto a site and get some work done, and then someone (hopefully) will come and live in the building. The building is not complete until it is occupied; which means, significantly, that it has been sold and found its place in the market or has the recognition by someone of its value as a home. In this simple illustration we can easily see two important things. Firstly, that the 'object' of the process becomes more and more what we might call *alive* or even *conscious*: that is what going up in level means. Secondly, that the total process goes in steps with critical moments of transition: there is, for instance the moment of 'breaking ground' and then the moment of signing the contract of sale or its equivalent, both of which have almost 'ritualistic' acknowledgments. There are *three phases*, each dominated by a different agency: first architect, then contractor

Figure 2 *Three Worlds of Building a House*

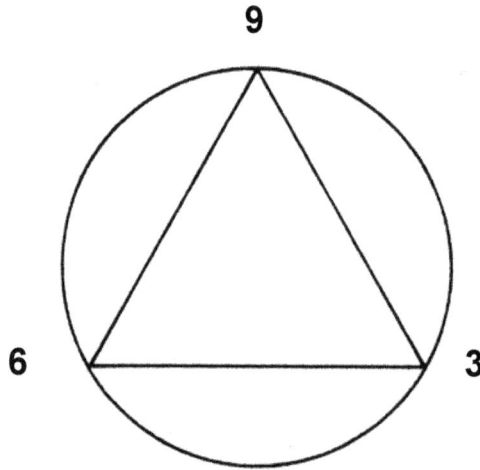

Figure 3 *The Primary Relationship or 'Logos'*

and finally house owner (see Figure 2). It is important to note that these three agencies need to communicate with each other, while it is notorious how often there are conflicts and misunderstandings.

So, what stands behind structured process is a relationship of three worlds or their representative agencies. This figures in the enneagram as the triangle, the sides indicating the three kinds of process and the points the critical moments or transitions (see Figure 3).

We use the term 'relationship' to emphasise that this not a mechanical process but a living and potentially conscious action. The word reminds us of people rather than machines. And people have an inward side and are not just to be pushed about or 'used'. Between people and machines—in terms of level—is the world of life which has its own rules and self- governance. There can be life not just in organisms but also in the way we work and this life requires autonomy.

To give a more sophisticated picture of the triplicity we can imagine the three processes going on simultaneously, and make a step diagram. From 0 to 3 there is one world, from 3 to 6 there are two, and from 6 to 9 there are three. Each of the three has its own history or origins and futures. The contractors go on to build other buildings, the home-owners maybe sell up and move to another location (as their children grow up, for example).

As the Figure 4 shows, there is not only simultaneity but sequence: we have to get sufficiently far with one process in order to relate it to another one; doing different processes at the *same time and place* is confusion. It is important, for example, that agency is 'handed over' in a clear manner.

Cybernetics and Learning

Cybernetics (from the Greek word for 'steering') introduced the all-important idea of *feedback*, a notion which still, apparently, very few politicians understand. The basic notion is that if one acts on a system then it will act back and *change whatever change one wanted to achieve*. In physics, for example, a physical system will behave such as to create a force that opposes whatever force is acting upon it. This is enshrined in Newton's Third Law. But, of course, politicians and reformers who want to 'do good' tend to ignore the realities they seek to control and persist in ineffectual and even harmful behavior. They do not 'listen' to what reality is telling them. By 'reality' here is meant simply the concrete situation as it is.

A craftsman will respond to the material with which he is working and will not impose a form that does not suit it, and will even *follow* its indications; as in carving wood. This illustrates another cybernetic principle enunciated by Gregory Bateson

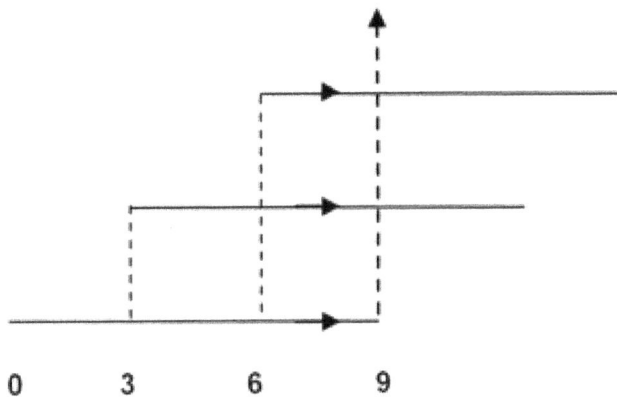

Figure 4 *Simultaneity of Triple Processes*

(1972) as *double-loop learning*. Single-loop learning is when one 'sticks to the plan': if I have decided to carve a man out of this piece of wood then a man it will be, and I apply myself to get that result. In double-loop learning one can have one's mind changed and realise that the piece of wood better suits being carved into a woman. In other words, I can respond to the realities I encounter and *change my objectives*. Bateson said, in his last lecture (1979):

> *I have offered you the idea that the viewing of the world in terms of things is a distortion supported by language, and that the correct view of the world is in terms of the dynamic relations which are the governors of growth.*

Single-loop learning was, notoriously, exhibited by the Soviet Five Year Plans; though of course when they did not work out there was no learning only an increase of false propaganda. Double-loop learning often features in fairy tales when the hero responds to someone he meets on his quest instead of, as his brothers do, ignoring them and rushing on. It is somewhat regularised in science in experimental method which seeks to test mental constructs against the evidence of physical reality (a still revolutionary idea in the western world when Francis Bacon first proposed it in the 17th century).

Though cybernetics was born from the world of engineering it rapidly became an exploration of human social reality. In the words of pioneer Gordon Pask, we are 'participant-observers'. We can picture that at first engineering was producing objects that impinged on the human social world and then *awakening to feedback from this world* to consider people as responsible for creating the visions out of which engineering springs (the word 'engineer' originally meant 'ingenious' which in its turn meant 'inspired' or 'full of spirit'). This entailed that engineering take account of how people experience and represent reality and themselves. Actually, it required a new type of engineer who has been slow in evolving. The theory was called 'second-order cybernetics' as exemplified in this passage from pioneer thinker Heinz von Foerster (2003):

> *...a brain is required to write a theory of a brain. From this follows that a theory of the brain, that has any aspirations for completeness, has to account for the writing of this theory. And even more fascinating, the writer of this theory has to account for her or himself. Translated into the domain of cybernetics; the cybernetician, by entering his own domain, has to account for his or her own activity. Cybernetics then becomes cybernetics of cybernetics, or second-order cybernetics.*

This 'modern thinking' which may appear complicated, has its antecedents in ancient cultures, when people would call on a spiritual world to guide them. An icon painter would fast and pray to enable him to be in the *right state* to create the work. This is, in enneagram terms, making a relationship with the 'third world'. Not only do I have to be able to respond to what I am working with but I also have to learn how to make my response a 'good' one. The word is used advisedly. It can be taken as merely subjective or in even a Platonic sense of *The Good*. The three phases have characters that can be summarized as:

1. Directed, intentional   single-loop;
2. Realistic, responsive   double-loop;
3. Good, helpful to the world   triple-loop.

People who believe they know what is good from the start are liable to turn into tyrants and cause much harm. It is from being able to respond to realities—which means to change one's mind—that the possibility of goodness arises.

Our mixed discourse of morality and engineering may appear confused, but their conjunction here is deliberate and important. Value-added process involves *values* and values enter through and for people. Along similar lines we must remark that making money can be a phase 1 process and never come to take account of the effect it has on the systems involved (in phase 2) let alone acknowledge the real needs of society (usually hidden under the words 'economic system'). The enneagram can serve as a reminder of what a total or 'complete' process entails. Nearly all design and management systems are one-dimensional (phase 1). Here is not the space to examine exceptions but we can cite the Russian methodology of creative innovation TRIZ (Salamantov, 1999) and the educational methods of Edward Matchett (2010).

As many people realise, coming up with good ideas is relatively easy. The hard part is getting any one of them to work in practice. For all that, organizations continue to dream of 'assured solutions' and pay consultants high fees to come up with them.

THE THREE AND THE SIX

The transition from phase one to phase two is relatively speaking as from abstract to concrete. Phase one, typically, is having the idea, planning, calculating and so on. However expertly done it is not the real thing. Nobody can be sure something will work until they have tried it! Or, as is often said, the plan of battle goes out the window with the first skirmish. And yet the plan is necessary. What it does is to set us off in a certain direction, launch us on our path.

The transition from phase two to phase three can be overlooked. Take the case of writing a book. An author can conceive the idea, plan out the work, apply himself, get the text edited and ready for printing and so on; but the meaning of the book is *in its being read*. In other words, it depends on something outside of the author's control (there are cases of unethical authors writing positive reviews of their own work!). Again, staying with this example, the moment of 'letting go' can be quite traumatic.

In broad terms, the third phase is the bigger world, the public, the market, society and that sort of thing. It can also be the longer term welfare of an organization, beyond its current performance and productivity: is it becoming capable of learning from its environment?

However, we indicated that the three component processes are not only sequential but simultaneous, which means that they are accessible throughout. This gives the potential of a different way of working. In terms of individual agencies or people, it requires at least one person to be in touch with all three all the time.

Figure 5 shows a rather old-fashioned example of the necessary triadic relationship. Historically, the three were often at loggerheads or the one trying to dominate over the other. In some modern organizations such roles are now obsolete or merged. An important feature of some of these changes is that the organization (on its left hand side or third phase) is *involved* with its customers who now play a significant role in what is produced.

Such organizations tend to positively relate to the intelligence of their customers in contrast with the more traditional kind that tended to rely on them being kept in ignorance. This point is raised as an example of how the basic triad governing the process can be an *open* and not a closed system.

However it is incarnated, the simultaneous working of the three components means another kind of pathway may be executed. In Figure 6 of the enneagram this is portrayed in the cyclic hexad of the inner lines (our reference point in what follows is the small black circle at point 1).

Our imagined 'free spirit' has carte blanche to weave in and out of the processes as he pleases. The geometrical figure has an interesting mathematical basis but what is important is the visualization of this alternative *other way* of doing things. Actually,

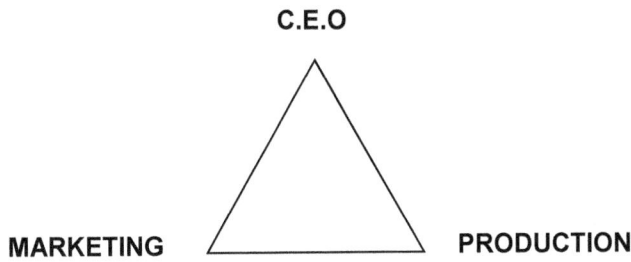

Figure 5 *Three Worlds of Business*

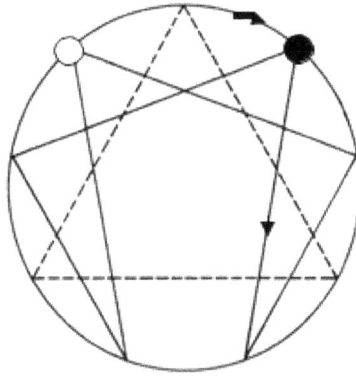

Figure 6 *The Primary Relationship or 'Logos'*

the arts tend to exemplify this way. For example, take a composer who gets an idea (black circle), but then immediately sits down and tries things out on his piano before he goes any further. He wants to know *how it might sound*. Next there is something that might sound a bit crazy: he 'listens' in another way to what the work *will be*. This takes him right over to the other side of the enneagram (see the small white circle at point 8). From there he comes back to the requirements to be made of the piece.

We said that it might sound crazy—because we are not used to consider operating from what we hope to achieve and working, so it seems, *backwards*. However in business circles a model proposed for entrepreneurs is given the title the *Merlin factor*, after the myth of the magician Merlin who lived his life backwards, from the future. Charles Smith (1994) avers:

> It begins with a personal quest to cast off the shackles of old habits of thought in order
> to reinvent the future. It takes hold in the present through the effort to enroll others
> as committed participants in the enactment of a new collective purpose. It gathers
> momentum with each 'impossible' obstacle that is overcome. The essence of the Merlin
> Factor process in organizational leadership is simply stated: what you choose for your
> future is more important than what you know about your past or present capabilities.

The hexadic cycle gives another meaning to the word *structure* in 'structured process'. In effect it raises the possibility of dialogue amongst all those involved *as equals*. The genius it reveals does not have to be that of some exceptional person but is within ourselves and accessible if we can listen to each other. That is why such approaches as Richard Knowles's workshops are solid contributions to making the processes they address *work well*. The higher order cybernetics is in our relationships, demanding honesty, trust and courage.

REFERENCES

Bateson, G. (2000). "Last lecture (1979)," in R.E. Donaldson (ed.), *A Sacred Unity: Further Steps to an Ecology of Mind*, ISBN 9780062501004.

Bateson, G. (2000). *Steps Towards an Ecology of Mind*, ISBN 9780226039053.

Blake, A. (1996). *The Intelligent Enneagram*, ISBN 9781570622137.

Knowles, R. (2002). *The Leadership Dance*, ISBN 9780972120401.

Matchett, E. (2010). *Fundamental Design Method*, ISBN 9781906769185.

Pask, G. (1975). *The Cybernetics of Human Learning and Performance*, ISBN 9780091194901.

Salamatov, Y. (1999). *TRIZ: The Right Solution at the Right Time*, ISBN 9789080468016.

Smith, C.E. (1994). "The Merlin Factor: Leadership and strategic intent," Business Strategy Review, ISSN 0955-6419, 5(1): 67-83, http://www2.jenniferguy.com/merlin20f.pdf.

von Foerster, H. (2003). *Understanding Understanding: Essays on Cybernetics and Cognition*, ISBN 9780387953922.

Chapter 2

OLD WISDOM FOR A NEW WORLD IN CRISIS? THE ENNEAGRAMMATIC STRUCTURE OF INTEGRATED, OPTIMAL AND SUSTAINABLE PROBLEM-SOLVING

Cameron Richards

The concept of 'wicked problems' (e.g., Kolko, 2012) generally refers to problems and also organizational as well as public policy challenges which may be complex and without or resistant to simple convenient solutions. This paper outlines a framework for addressing all kinds of wicked problems which adapts the ancient wisdom of the enneagram as a basis for a theory and practice of integrated, optimal and sustainable problem-solving. In similar fashion to the insights of complexity/fractal/convergent/chaos theories of science and related models which bridge natural and human knowledge systems and thought, such a framework inevitably (we think) needs to be built on principles of convergent thinking, the emergent 'self-organizing' aspects of nature, and a dynamic yet interdependent view of interaction with or adaptation to any complex and changing economic as well as natural environment. The ideas developed in this paper are consistent with and also build upon aspects of Richard Knowles's practical and accessible Process Enneagram model.

INTRODUCTION

Evidence of how difficult decision-making has become is everywhere. Repeated modification of strategies has become a requirement for firms and the rapid adjustment of public policies has become the norm for governments... when organizational plans, corporate strategies or national policies don't work, there is often a reflex suspicion that leaders are either incompetent or corrupt...

D. Rycroft & R. Kash, *The Complexity Challenge*, p. 17

As our conditions and environmental context keeps changing, the balance keeps shifting. It's the Leadership Dance

R. Knowles, *The Process Enneagram: A Tool from The Leadership Dance*, Part 2, p.7

The renewed application of an ancient symbol of transformation (the enneagram) by Knowles (2002) to support the 'leadership dance' needed for 21st Century organizational adaptation to a fast-changing world in crisis also exemplifies the timeless application of and convergence between ancient human wisdom and unprecedented 21st Century challenges. As J.G. Bennett (1974) outlined in his studies based on the work of Gurdijieff and Ouspenksy, the enneagram derives from ancient knowledge traditions which have yet long converged with the foundations of modern knowledge. This is exemplified, for instance, by how the seminal Pythagorean influences on the emergence of Western thought and mathematics were in turn based on associated notions of 'sacred geometry', musical harmonics, and philosophical as well as aesthetic principles of the golden ratio or proportion—a symbol for natural growth as well as symmetrical proportions in time and space alternately expressed as the Fibonacci numbers, the logarithmic spiral, and the phi geometrical proportion. The natural spiral structure of the golden ratio or section makes it an exemplary symbol of the link between internal and external aspects of systemic transformation (Livio, 2002). In contrast, the enneagram is rather the exemplary process symbol to describe a 'self-organizing systems' view in time for emergent, strategic and intentional outcomes-based problem-solving. This is especially so in relation to human organizations needing to adapt to increasingly complex and fast-changing environments. In this way it might be proposed that the enneagram process retains its exemplary relevance for a 21st Century world in apparent perpetual crisis.

As Bennett further points out, the enneagram structure encourages as well as reflects a dynamic 'triadic' or triangular perspective which represents a remedy for rigidly linear, hierarchical, and either-or (i.e., dualistic) notions of thinking, planning and acting (p.2). Rather it projects various notions of human development as an emergent and convergent knowledge-building process as well as dynamic interaction or dialogue in time and space. Knowles's distinction between 'machine' and 'living systems' applications of the enneagram process corresponds to a related delineation between negative and positive cycles of knowledge transformation. Knowles's conception and application of the enneagram process thus focuses on how human organizations are complex adaptive systems reflecting the self-organizing, emergent, and interdependent principles which are basic to complexity science and related models (fractals, chaos theory, and so on). Prigogine's concept of dissipative structures has been pivotal in recognizing the function of self-organizing order in nature not just out of chaos but also apparent 'emptiness'. As Jay Forrester, Niklas Luhmann, Mario Bunge and others have usefully outlined, systems thinking and models of transformation are an antidote to mechanistic, reductionist, and positivist models of knowledge. Likewise they are also applicable across realms of material formation, biological growth and social orders (including the animal kingdom and its evolution) as well as human knowledge systems reflect the crucial problem-solving importance of constructive, multi-disciplinary thinking (Klein, 2006). In this way many commentators have recognized how human organizations function as naturally complex adaptive systems of energy and information in relation to changing environments (e.g., Mitleton-Kelly, 2003).

The paper will explore the exemplary application of the enneagram process for integrated, optimal and sustainable problem-solving in terms of three related process stages. The first stage will link the natural human capacity for identifying and addressing problems to a systems view of the transformative interdependence of different kinds of knowledge. This stage will be explored in relation to the fundamental insight of the enneagram model that all specific aspects of life or 'unity/agency/organization' are linked in terms of a common or 'deep' universal structure. The second stage of discussion will then explore how the enneagram's triadic (3-6-9) principle also exemplifies a globally dynamic and strategic framework for overcoming the self-defeating aspects of either-or thinking, of such dualistic notions as *theoretical vs. practical* (or *pure vs. applied*) knowledge, and *top-down vs. ad hoc* approaches to policy-making, planning and general 'problem-solving within various *public vs. private* organizational as well as applied contexts. The enneagram's hexad process prescription (the underlying 1-4-2-8-5-7 formula of 'unity' as process and not just a function or form) for linear thinking and hierarchical organization is thus further discussed in the third stage in terms of several related concepts which epitomize integrated, optimal, and sustainable problem-solving.

THE UNITY OF ONE (OR MORE) AS SYMBOL FOR THE INHER-ENT UNIVERSALITY OF SPECIFIC SYSTEMS

We're trying to deal with a whole array of integrated problems—climate change, energy, biodiversity loss, poverty alleviation, and the need to grow enough food to feed the planet—separately. The poverty-fighters resent the climate-change folks; climate folks hold summits without reference to biodiversity; the food advocates resist the biodiversity protectors... In short, we need to make sure that our policy solutions are as integrated as nature itself. And the only way to do that is re-immerse ourselves in systemic thinking

T. Friedman (2008), *Hot, Flat and Crowded*, Picador, p. 224

We may be intelligent beings, but in truth, so far as the world process [enneagram] is concerned, we are just half-cooked food.

J.G. Bennett, *Enneagram Studies*, p. 11

The enneagram symbol is composed of three distinct yet inter-related parts of a convergent process—a circle, triangle and a hexad arrangement with (like the related pentagon and also pentagram symbols) a lateral rather than radial symmetry. The circle is naturally a universal human symbol of unity and related notions of integrity as well as also order out of apparent nothingness or emptiness. Also reflecting the potential links between formal mathematical equations and universal *vs.* specific notions of reality, 'unity' can either be represented as 0 or 1. In the overall enneagram the circle also represents a cyclic sequence. The ennead sequence 1-9 is typically viewed in terms of an inward arc from 1-4 giving way to an outward arc of 5-8 with the initial point 9 symbolizing the culmination of a complete cycle of unity as a process of transformation. The final number of the base 10 number system, 9 simultaneously represents a restoration of unity but also a process of irreversible transformation.

The Neoplatonist Iamblichus left a record of the related philosophical as well as functional importance for the Pythagoreans of 'the first ten numbers' which they represented in the *tetractys* symbol of the musical, arithmetic and geometric ratios upon which the universe was thought to be built. Likely the original prototype of the enneagram, the tetractys also naturally frames a regular hexagon. Pythagorean mathematics are not only a key foundation of Western thought linked to Plato (e.g., in terms of the Platonic solids) but also the exemplar of a global tradition which gave rise to decimal fractions, algebra, and eventually algorithmic reasoning. As Iamblichus (1998: 105) put it around 350 CE, 'the ennead is the greatest of the numbers within the decad and is an unsurpassable limit... everything circles around within it, it is clear from the so-called recurrences: there is progression up to it, but after it there is repetition'. In such ways the Pythagorean 'law of the octave' has been a direct inspiration for seminal models in modern science such as Newland's discovery of the periodic table of chemical elements, Tesla's revolutionary invention of the alternating current generator, and Penrose's pentagonal model of non-periodic tiling and its discovered application (thought impossible) to the natural growth of quasi crystals. The Pythagorean ennead has been also linked to various other ancient world models—especially in relation to what Karl Jaspers appropriately called the 'axial age'. As well as also linking back to the ancient Babylonian, Phoenican, and Egyptian formulations of number, knowledge, and mythology (e.g., West, 1993), this model has been traced as a key influence on later Neoplatonist, Gnostic, Kabbalistic and Sufi knowledge systems (e.g., Aczel, 2000). As West (1993: 56) concludes, 'whereas the tetractys shows the Grand Ennead made manifest, the enneagram shows it in action as Seven, the octave, number of growth and process, interpenetrating Three, the basic triune nature of unity'.

So what has this really to do with practical human problem-solving today? Reflecting the generic definition of a problem as "a perceived gap between the existing state and a desired state, or a deviation from a norm, standard or status quo" (www.businessdictionary.com), from a systemic perspective all human problems are 'complex' even if apparently or actually 'simple'. Simple problems (e.g., a bacterial infection, a clogged up fuel filter, or a personality clash within a business organization) which may initially seem more serious might well be quickly addressed and efficiently resolved. However good doctors, mechanics, and leaders all know that both simple and complex problems are all ultimately about restoring the natural and deep-level efficiency or health of a particular 'system' whether this be a patient, a car or a business organization. As the wicked problem concept illustrates, the world of actual human experience and organization as well as all nature generally is ultimately and intrinsically complex, interdependent, and open to perpetual change. Superficially 'simple' problems ever conceal a latent complexity, yet ostensibly 'complex' problems are ultimately quite simple in principle. However people continue to impose and apply both top-down or rational and ad hoc or heuristic approaches to human 'problems' as if they can always be simplified, pin-pointed, and resolved in a de-contextualized vacuum. Conversely we also still tend to view individuals and groups as fixed identities when all human systems are also endlessly both parts and wholes above and below any axis of unity. Above all else then, as a symbol of the ancient wisdom process the enneagram provides a framework for not only appreciating rather the internal or transformative aspects but how this might better inform and support the external or translational aspects.

Following the influence of John Von Neumann in particular, modern knowledge systems have tended to focus on *problem-solving* in terms of abstract mathematical equations and *decision-making* in terms of the economics of risk and probability (cf. also Kahneman, 2011). Yet if unity is viewed as a systemic entity or agency in time and space then it is both internally unique but externally inter-connected if not interdependent in the manner of 'network' organization. It can be reasonably argued that the most effective modern scientists (as well as leaders) typically organise their ideas and inquiry around relevant focus problems—and not just (top-down) rational abstraction or (ad hoc) empirical descriptions in isolation. So too in every phase of human cultural history people have adapted to new or changing economic as well as natural environments (e.g., hunter-gatherer, agricultural, urban-industrial, and global knowledge society) with a basic problem-solving orientation of knowledge construction and learning. As the scientist and writer Jared Diamond (2005) has pointed out, every level of human society is ever potentially innovative, sustainable, and socially relevant—or open to potential paralysis and 'collapse' if not. This might be appreciated in terms of the useful distinction made by John Biggs and others between surface learning and deep learning in education. That is, we might better recognize the deep knowledge-building convergence between the emergent process of 'knowledge as applied understanding' as distinct from the specific surface-level descriptive information and technical or skills efficiency of 'knowledge as explanation'—the alternate or complementary aspects of what we have elsewhere referred to as 'global knowledge convergence' (Richards, 2011b).

As Figure 1 outlines, the enneagram process represents an exemplary symbol of how any self-organizing system or human process inevitably involves three key axes of organization in time and space. It does so above all else in terms of the how the 1-4 sequence typically represents an internal arc or sub-system of an overall transformation process, whereas the 5-8 sequence likewise is typically conceived as an external or applied focus of change and interaction. As Knowles (2003: 121) outlines, the 1-4 sequence is also about creating value whereas the 5-8 sequence is about realizing value. As further depicted in the enneagram depiction on the left, any general process of transformation is typically conceived as an enduring identity or structure in time and space. The diagram on the right-hand side converts this into a *systems* view of the human knowledge-building process—a *micro-macro* interplay of both specific and universal elements in any human or other knowledge system. In this view also, a particular system of either individual or group identity also incorporates both an *internal axis* of formation and integration and *external axis* of emergence through interaction with a changing environment. In other words, systemic integration, optimization, and

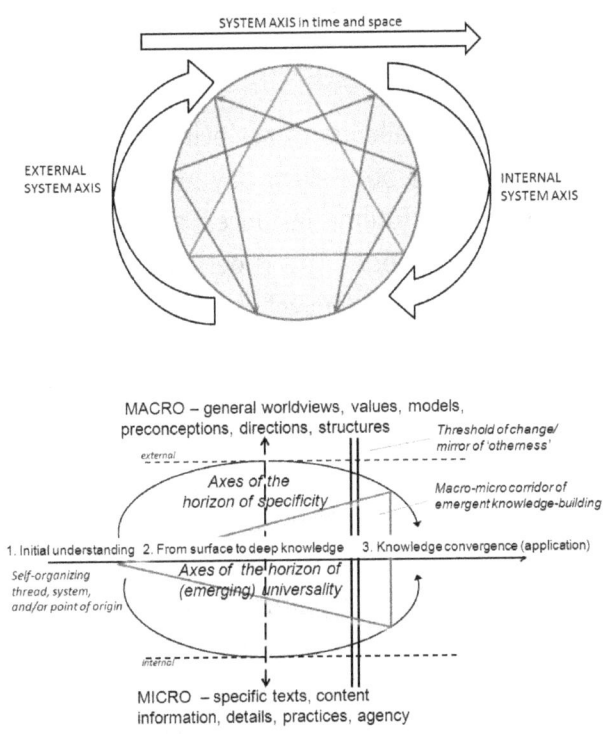

Figure 1 *The enneagram and the convergent axes of 'unity' which inform human knowledge-building. Bottom figure adapted from Richards (2012a).*

sustainability build on an *alignment of axes* within as well as between different systems—which may include individual agents as well as social networks of organization. This diagram also adapts Ricoeur's (1994) insight that the inevitable interplay of specificity and universality as well as 'sedimentation and innovation' in human knowledge should be viewed in terms of the transformative rather than 'either-or' or dualistic perspective of: (a) an initial or *naïve understanding* contrasted with (b) a provisional phase of *critical explanation* giving way to (c) *applied understanding* grounded in the deep not just surface convergences in any knowledge system.

The representation in Figure 1 of the enneagram as an exemplary process structure or symbol of human problem-solving builds on the adaptations of Bennett and Knowles. Bennett typically used a food metaphor of meal preparation to depict and explain the enneagram process as a systemic transformation typically resulting in some kind of convergent outcome or product. Knowles's unique and effective adaptation of the enneagram process to depict human organizations typically focuses on how this might be applied to overcome the negative cycle and either-or thinking of 'the management trap' described by Argyris and others. He thus refers to the dynamic interplay between the internal function of 'operational management' and the external or overall direction of 'strategic leadership' in order to achieve coherent efficiency and emergent effectiveness in the self-organizing capacity of any social network or institution.

Our related focus on the challenge of achieving integrated, optimal, and sustainable problem-solving across both every-day practice and high-level strategizing (as well as both simple and systemic instances) embraces and builds upon these adaptations of the enneagram model. This is perhaps best exemplified even today by Socrates's elenchus method of linking yet also distinguishing between specificity and universality in human thought—a recognized prototype of the modern scientific method as well as the ideas of both Plato and Aristotle (Scott, 2002). The key to this method for also addressing difficult problems (aporia) is the strategic use of relevant questions or related threads of inquiry (elenchai) to assist both simple and complex human problem-solving by encouraging focused thinking, collaborative dialogue, and open-ended inquiry based on open-minded humility not close-minded arrogance or condescension towards others. It is well-known that Socrates was a self-conceived 'mid-wife' helping others to think and achieve deep-level insights and understanding. He based his convergent model of balancing *applied critical thinking* and also (like Gurdijieff) *emergent self knowledge* on a fundamental recognition that the 'wise ignorance' of those who openly acknowledge the gap between 'what they know and don't know' have a significant and potential future advantage over the 'arrogant ignorance' of those who 'don't know they don't know' (Richards, 2010).

Also corresponding to Knowles's related distinction between 'machine' and 'living systems' paradigms of the enneagram process, we compare the negative top-down vs. ad hoc cycle of eventual policy paralysis with a positive and sustainable framework which aligns the *internal axis of accountability* with the *external axis of feedback* to effect *systemic resilience with integrity.* This is contrast to 'parasitic resilience' where another system, organization or agent is hi-jacked or colonized for its energy and other resources. Whereas accountability without feedback is top-down rationalism in conflict with ad hoc effort, feedback without accountability results in information overload and retrospective justifications based on arbitrary selections of evidence. A negative cycle is not sustainable and will result in eventual policy or leadership paralysis and inevitable failure. In this way also we might discuss the global need to align and converge the distinct notions of *ethics* as an internal and modern imperative of personal conscience on one hand, and on the other a traditional moral focus on community standards reciprocally linked to social networks (Richards, 2012c).

In similar fashion to the selective *evidence-based rationalizations* of much scientific theorizing as well as political decision-making, we might further distinguish between the negative cycle of 'downwards and backwards' thinking and planning in contrast to the 'upwards and forwards' (or emergent and convergent) approach of integrated, optimal and sustainable outcomes-based problem-solving. To the extent that the 'upwards and forwards' approach *turns on its head* retrospective justifications or merely hopeful anticipation, it exemplifies how the influential and transformative GROW (Goal-Reality-Obstacles-Options-Way Forward) model in sports and other coaching is based on the notion of aligning external interactions with what Timothy Gallwey calls the 'inner game'.

THE RELATION BETWEEN COMPLEX PROBLEMS AND THE ENNEAGRAM'S DEPICTION OF THE 'LAW OF THREE'

Complexity suggests that to survive and thrive an entity needs to explore its space of possibility and to generate variety... that the search for a single 'optimum' strategy may neither be possible or desirable

E. Mitleton-Kelly, *Complex Systems and Evolutionary Perspectives on Organizations*,
p.14

For a man who is able to make use of it, the enneagram makes books and libraries entirely unnecessary... a man may be quite alone in the desert and he can trace the enneagram in the sand and in it read the eternal laws of the universe. And every time he can learn something new, something he did not know before

G. Gurdijieff, quoted in P. Ouspensky's *In Search of the Miraculous*.

The triangular section of the enneagram symbol depicts what Bennett calls the three 'sources' of any dynamic system process—the common resources also of Knowles's model of Self-Organizing Leadership. A related rhetorical 'law of three' was first mentioned in Aristotle's *Rhetoric*. In contrast to the transformational process of the hexad sequence discussed further below, the triadic 3-6-9 structure of the enneagram represents the threefold foundation for as well as critical stages of any overall transformation process. As Gurdjieff (2004) himself put it in "The fundamental laws of the 3 and of the 7, of the active, passive, and neutral principles, the laws of activity, are to be found and confirmed in everything, and therefore [the enneagram represents]... a knowledge of the world's structure".

As a remedy for dualistic either-or thinking and also top-down vs. ad hoc planning, a triadic or dialogical perspective can be applied in two distinct but related ways which also depict an alternately external and internal perspective. Firstly there is the surface level of perceived opposites or contrasting terms of external and arbitrary differences (such as male vs. female gender). Secondly there is a deep level notion that in time a meeting or contrast of opposites might produce a new emergent output or convergent reconciliation of some kind extending from natural or scientific oppositions (e.g., negative vs. positive charge) to cultural ones (e.g., the traditional Chinese principle of 'yin and yang' interdependence). As Ricoeur proposes, such an emergent connection typically reflects three progressive stages of knowledge-building proceeding from *naïve* to *critical* to *applied* understanding and explanation. Thus a constructive policy-building approach to problem-solving should integrate desired outcomes, anticipated audience/stakeholders, and also aspects of local context as a foundation for transforming information and perceptions into knowledge. Kahneman (2011) generally applies a typically dualistic as well as statistical approach to the distinction between 'fast and slow' (i.e., intuitive/associative vs. logical/reflective) modes of human thinking. Nonetheless he had to concede that the 'convergent thinking' of a balanced optimism can lead to a more effective, resilient, and coherent implementation of plans or ideas which in turn help create a positive self-fulfilling prophecy of success able to defy unlikely and even apparently 'impossible' odds (p. 263).

As Knowles also suggests with his 'leadership dance' remedy for the 'management trap', this might be usefully discussed in relation to the overlapping models of Chris Argyris and Donald Schön across three progressive levels of application. Schön's particular concept of 'reflective practice' represents a dynamic and strategic resolution to the typical mind-body dualism of *thinking vs. doing* (also theory vs. practice, rationalism vs. empiricism, top-down vs. ad hoc, etc.)—or rather the alternate negative link between 'thinking without doing' and 'doing without thinking'. Whilst this notion of action learning or performance is specifically focused on concrete levels of individual practice, the distinction between *single-loop learning* and *double-loop learning* collaboratively articulated by Argyris and Schön corresponds in practice to that between surface and deep learning indicated earlier. Like Kurt Lewin's related 'unfreezing-refreezing' model of change management, the characteristic 'double-loop learning' strategy of transforming deep organizing structures of thought, knowledge, and capacity can work well in a stable environment but is perhaps not so effective in practice to optimally and sustainably deal with constant change. However it at least points to the notion of a common deep level of convergence involving an internal as well as external axis of thought, activity, and strategizing.

A third level convergence has been perhaps most effectively outlined in Schön's final work (with Rein) which focuses on the design challenge of re-framing in order to achieve problem-solving resolutions to 'intractable policy controversies'. This work concentrates on how policy problem-solving strategies within any organizational context requires integrated and convergent collaboration between any or all the four macro-stakeholders (government, private sector, academia/intelligentsia, and the wider community or civil society). In related fashion to Socrates's elenchus method for resolving 'aporia' through engaging a 'thread of inquiry' set in motion by a relevant question, Schön and Rein's (1994) proposal that frame reflection and transformation is a crucial key to productively tackling and often overcoming seemingly impossible problems invokes our support here for the idea that the enneagram exemplifies a deep foundational structure of 'integrated, optimal and sustainable problem-solving'.

Such an approach involves an emergent *outcomes-based* rather than retrospective or rationalist *evidence-based* inquiry and problem-solving. In terms of the enneagram process it adopts the constructive version of the 'law of three' to outline a practical example of formulating a framework for addressing 'wicked problems'. The initial phase involves achieving a provisional or working foundation. On this basis the second phase seeks to prioritize the various relevant internal and external factors or contributing problems. Following on from or simultaneously to this, a third phase seeks to develop an emergent and convergent solution. The implied strategy then is to 'optimize' the problem-solving process in terms of transforming any relevant data and information into applied knowledge and understanding. This might be appreciated in terms of recognizing the interplay of internal and external axes of inquiry which together constitute the so-called *data-information-knowledge wisdom pyramid* (see Figure 2) used in such areas as 'management information systems' (e.g., Fricke, 2009).

In such applications 'wisdom' is typically seen as unknowable or referred to only ironically. The accumulation and description tendencies of an external axis of empirical data and organized/rationalized information is redeemed or open to be transformed in terms of some focus outcome in relation to an internal axis of knowledge, experience and understanding. In this way 'wisdom' need not be an accidental by-product or outcome of accumulation and complexity but actually a deep foundational process based on the quality of experience and understanding not quantity of information

Specific re-solution,
universal application
9

An initial 'thread' of purpose or
inquiry *(self-organization/
innovation)*

8 Sustainable convergence/
dynamic equilibrium
[process]

1

7

Surface vs. deep
coherence/integrity

2

*Corridor of
emergence*

Optimal
emergence/growth
[structure/form]

Integrated/
systemic approach
[function]

Develop an 'ecology of
knowledge' *[internal/
communication axis]*

Adaptation to complex &
changing environments
*[external/knowledge-building
axis]*

6

3

Threshold of change

5

4

---

*communication/knowledge as
a process of translation*

*communication/knowledge as a
process of transformation*

LOCAL

fixed/passive/
objective

particular/
individual

Data

description

microcosmic vs hierarchical

1. SPECIFIC
(differences/ variations)

Information

surface

2. GENERIC FORMS
(organizing patterns)
process

experience

SELF/US
(same)

OTHER/THEM
(difference)

deep

Knowledge

understanding

3. PRE-GENERIC
'SUBSTANCE'
(universal patterns)
egalitarian/dialogical

*hierarchical human
knowledge and social
organization*

Wisdom

dynamic/interactive/
ever-changing

general/collective

GLOBAL

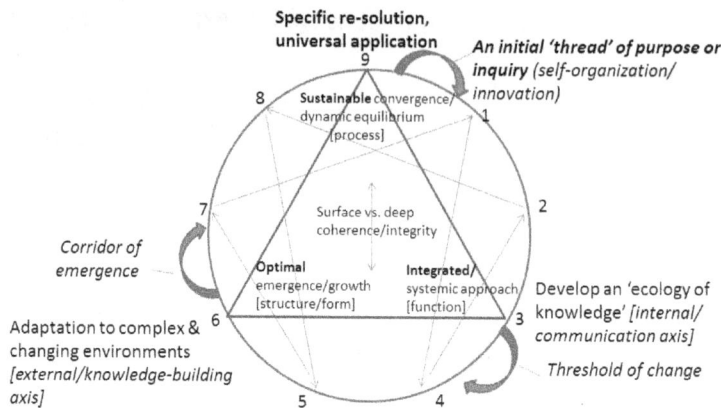

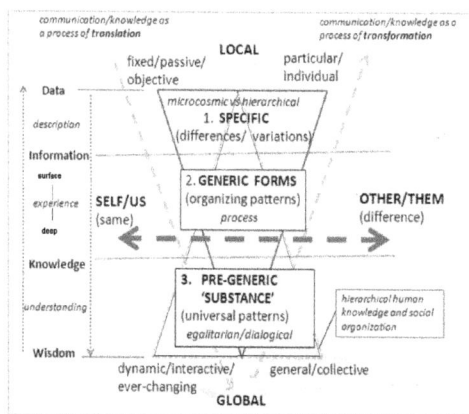

Figure 2 *The enneagrammatic generators of the 'data-information-knowledge-
wisdom pyramid'. Bottom figure adapted from Richards (2012d).*

(Richards, 2011). In relation to his workshops on Self-Organizing Leadership in organizations, Knowles likes to point out how the Process Enneagram makes this process 'smooth and easy… it just seems to happen' (personal correspondence).

Bennett has outlined how the enneagram represents an exemplary 'cycle of transformation' in terms of three critical phases which link the conventional 1-9 stages with the 'integrated, optimal and sustainable' hexad formula of 1-4-2-8-5-7. The critical points of 3 6-9 thus locate where the most effective teachers, leaders, and coaches can make macro (or outcomes-focused) or even micro (i.e., 'carrots and sticks') *interventions* to encourage the process of emergence and convergence. As a process, then, the interval or transition between the sequence points 9-1, 3-4 and 6-7 represent particular stages of potential transformation where the inner 3-6-9 triadic structure of transformation links to not only the ostensibly linear and hierarchical 1-9 sequence but also the 1-4-2-8-5-7 transformative sequence. The systemic process is conceived at the 9-1 point transition in relation to how a new cycle represents a particular self-organizing process of problem-solving. Just as it is specific and unique within a particular space and time it also universally or systemically constitutes a kind of transfer and transformation of the *culmination* of existing or prior knowledge.

As outlined in Figure 2 the internal arc is about the transformation of data and information into a 'knowledge ecology' (or eco-system) as a whole-parts synthesis which overcomes obstacles of change in time at the 3-4 transition. Organizational as well as individual learners might always be better assisted or even challenged to more actively negotiate what, following the work of such people as Victor Turner, Joseph Campbell, and Eric Erickson, we might refer to as the intrinsic *threshold of change*—which if successfully negotiated is ever a provisional (and if not, a perpetual) stage of frustration, confusion and uncertainty or indeterminacy. In terms of organizations or networks this will require an effective communication of stakeholders sufficiently working together, as well as agents who become reflective practitioners. The associated knowledge ecology developed should provide the foundation also for a related 'external axis' of emergence and convergence as an adaptation to complex and changing environments. This follow-on stage is especially negotiated at the 5-6 transition as a process of development or even growth within a particular *corridor of emergence*. This process can be assisted as an integrated and dynamic interplay of macro directions and micro interventions in the unfolding of some strategy or plan to support or guide its actualization or achievement. In this view the challenge of 9 is to try ensure or mediate the convergence between *internal systemic growth* and equilibrium or homeostasis in relation to *ever-changing external conditions* or environments. The diagram above thus adapts a complex problem-solving process model to build on Knowles's conception that the 3-6-9 transitional point represent aspects of relationships, information and identity in the Self-Organizing Leadership of various human organizations.

The enneagram process does not just have a natural superiority over 'books and libraries' as Gurdjieff suggests, but arguably also the most advanced computers or related functions of artificial intelligence and algorithmic calculation. This is because it involves a universal formula which engages, frames, and features the internal or self-organizing aspects of human knowledge as well as material formation, biological evolution and social organization. As suggested earlier, nature tends to follow an innate trajectory of adaptation to changing physical environments which recapitulates the balanced of internal and external forces (also form and function) in turn epitomized by the innate spiral rather than hierarchical or linear structure of the golden mean or section. Social Darwinism as well as the theory of natural selection in physical evolution tends to view this process in a backward and downwards perspective as either determined and inevitable or simply ad hoc and accidental. But an upwards and forwards view of the related processes of decision-making, strategic planning and applied problem-solving recognizes that the naturally slow and dialogical emergence of an 'optimal solution' to any dilemma, predicament or obstacle can be accelerated as a structural habit or generic (and indeed genetic) disposition and not just a unique one-off event. The key to this process is to recognise the importance of achieving a specific direction, thread or purpose and then adapting and refining and building on this in terms of application to or implementation within a unique local context. In this way we might talk about designing or preparing an accelerated 'corridor of emergence' as foundation for regularly, effectively and sustainably achieving 'optimal solutions' to any simple or complex problem.

THE ENNEAGRAMMATIC FORMULA FOR INTEGRATED, OPTIMAL, AND SUSTAINABLE PROBLEM-SOLVING

The problems—poverty, sustainability, equality, and health and wellness—that plague our cities and our world... touch each and every one of us. These problems can be mitigated through the process of design ... Due to the system qualities of these large problems, knowledge of science, economics, statistics, technology, medicine, politics, and more are necessary for effective change. This demands interdisciplinary collaboration, and most importantly, perseverance.

J. Kolko, *Wicked Problems: Problems Worth Solving*, Section 1

The enneagram symbol is directly relevant to the scientific procedure... scientific experiments are bound by the same qualitative laws, structural principles, modes of patterning—however they may be called—as any other situation in which the presence of completing processes can be discerned.

K. W. Pledge, *Structured Process in Scientific Experiments*, p. 84

Bennett has pointed out how the decimal system has reinforced the organizing principles of symbolism which make up the enneagram process: 'we obtain a symbolism for one as an endless recurrence of the number 9' (p. 2). That is, just as 1/3 = .333 recurring so too 3/3 = .999 recurring. In related fashion, the transformative sequence 1-4-2-8-5-7 corresponds to the decimal conversion of any seventh fraction of unity (i.e., 1/7-6/7). As indicated above, the enneagram's 1-4-2-8-5-7 formula of transformation is linked in particular with a Pythagorean view of the integral and not just sequential or cumulative functions of number. In this view the number 7 has a key role in an ennead model of order reflecting the harmonic principles associated with the intervals which make up the musical octave obtained by doubling or halving the rate of vibration. The enneagram's hexad formula thus builds upon the concept that a system or unity of order is most integrally exemplified by 'seven-level spectrum' models of human needs, organization and consciousness refined in recent times by such innovative thinkers as Arthur Young, Andrew Barrett, Don Beck, and Ken Wilbur (e.g., Barrett, 2006). However there is an apparent paradox about how the enneagram alternately reflects but also apparently disrupts or contradicts this. Whilst a linear and hierarchical view of the sequence of numbers and steps reflects a transformative set of levels or stages, the formula of the enneagram process represents the ostensible *zigzag* transformation of this pattern in a way which initially does not make sense. The enneagram's hexad formula really only begins to make sense when appreciated in terms of the related internal and external (or descending and ascending) axes and cycles of any overall self-organizing process.

If we were to outline as a set of stages and elements the basic process of systemic (i.e., integrated, optimal, and sustainable) problem-solving, then the 1-9 sequence described in Figure 3 represents a generally comprehensive and useful if 'downwards and backwards' overview perspective. However the 1-4-2-8-5-7 formula epitomizes the process of transformation itself and thus presents a rather 'upwards and forwards' approach to the 'threshold of change' and related generic structures of universal process. In effect it turns on its head a retrospective, descriptive, and merely 'linear and

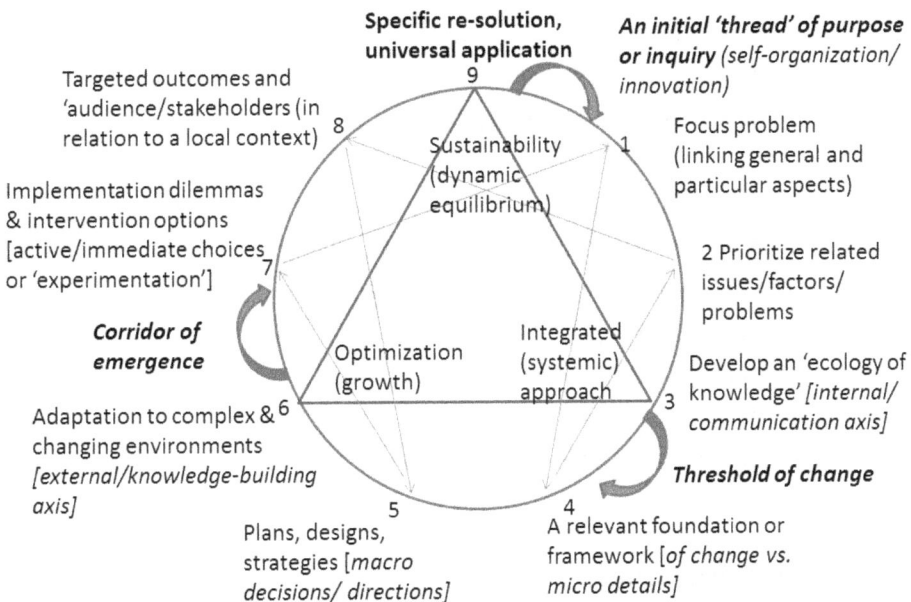

Figure 3 *The enneagrammatic formula of integrated, optimal, and sustainable problem-solving.*

hierarchical' view of human problem-solving and decision-making. It effectively does so by exemplifying the GROW principle of *focusing on and then constructively working back from a specific, reasonable, and appropriate target outcome* to align the axes of universality and specificity as well as the intrinsic and extrinsic aspects of motivation and knowledge-building in support of that desired outcome.

It is reasonably self-evident that a focused and constructive problem-solving perspective at sequence point 1 is likely to be more effective than a passive, vague, and merely hopeful approach to some issue or challenge. The logical step at point 2 is to aim to break down a focus problem or question into its associated factors, aspects, and related issues or problems. However it is also likely to be more effective in practice if this stage of the process is based on a foundation and framework which supports and informs the initial question or focus. Therefore the integrated approach needed for later optimization and sustainability suggests the transformation of a linear and retrospective as well as hierarchical perspective on the overall process.

The inward axis of a specific 'thread' of inquiry should simultaneously and inter-dependently link to an *external arc* in terms of an overall 'outcomes' reference point represented by the number 8. Thus in the same way the *1-4-2 transformation* represents how the learning process is never merely linear but a transformative jump across what Schön calls the 'practicum abyss' (i.e., the related content-process and thinking-practice divides) related to the external *8-5-7 transformation*. As Schön (1987: 13) it, the resulting 'knowing-in-action' from any 'practicum' demonstrates how there is "an art of problem framing, an art of implementation, and an art of improvisation—all necessary to mediate the use in practice of applied science and techniques". In contrast to how the problem-solving foundation of the internal axis is most effective when supported by a knowledge ecology making sense of the 'micro' detail of a local context, so conversely the external arc of designing and developing a solution to any targeted problem involves some macro direction (or 'vision of possibility') as a focus not just plan to support an emergent and sustainable systemic solution. Thus we outlined a related 'corridor of emergence' where a problem-solving process and thread of inquiry is able to anticipate future obstacles—yet also adapt to changing circumstances with flexibility in terms of options for supporting interventions.

In the final related example outlined below we propose to further link our complex problem-solving adaptation here of the enneagram model to even the most challenging, topical and globally relevant kinds of problems faced by whole industries, nations and the world at large. In related work (e.g., Richards & Padfield, in press) we have proposed that in relation to the typical global as well as 'wicked problems' which confront humanity a more convergent, interdependent, and collaborative relationship is needed between the key macro-stakeholders of public policy, financial markets, and international affairs—that is, government (or international) agencies, private or commercial sector organizations, academia and wider society. This is in the context that as Jeremy Rifkin and Paul Gilding have so effectively articulated, the world has seen a 'marketization' or commodification of all aspects of human life even in cultural as well as natural domains traditionally conceived as a natural commons or public/community 'good'. As Rycroft and Kash (2004: 4) put it, "In the realm of [the global 21st Century 'complexity challenge'], organizational networks have blurred the line between the public and private sectors and between company strategies and public policy". On this basis—and adapting a new, more convergent concept of particular organizational/ sector/industry hubs of complex problem-solving—we have outlined a frame work requiring interdisciplinary collaboration in terms of an exemplary focus on authentic solutions which include but go beyond different organizational contexts of shared knowledge-building.

Such an approach attempts to reconcile not only a diversity of human stakeholders but also a range of critical factors. These extend from the 'internal aspects' of stakeholder consensus and knowledge management to 'external aspects' of the various and even conflicting views of the role of science and technology (or applied knowledge) in the fragile interplay between economic imperatives of growth and environmental requirements of eventual equilibrium or sustainability. If not framed in a systems (and related 'knowledge ecology') perspective then the much-touted concept of 'sustainable development' remains largely a contradiction in terms typically used as a superficial rhetorical or advertising ploy embraced by both rich developed nations and poor developing ones. As Rist (2008: 46) puts it, "'development' which is always presented as a solution, is actually a problem (as well as creating problems)". A rather deep and meaningful use of the term to match the influential Brundtland Report definition of human sustainability (i.e.*'meets the needs of the present without compromising the ability of future generations to meet their own needs'*) thus also requires an alignment of external equilibrium and internal growth. In this way Figure 8 outlines the four critical elements of integrated problem-solving for optimal solutions and sustainable development relevant to any complex or wicked problem—including national industries, multinational corporations, and other kinds of global or local organization.

As also depicted in Figure 4, a sustainable problem-solving framework therefore involves four distinct aspects and requirements or elements of integrated problem-solving and policy-building reflecting corresponding modes of human knowledge: 1. (communication, consensus and inter-dependence of)*stakeholder perspectives; 2. knowledge management*(of organizational vs. niche/individual/local human resources and performance) 3. *science and technology innovations* (applied knowledge building as extension); and 4. *complex environmental adaptation*(to changing natural vs. socio-economic contexts in time). These aspects provide the focus for outcomes-based problem-solving geared towards the 'optimization' of natural and human resources, an innovative as well as green approach to new science and technology solutions, and the process of achieving a foundation for sustainable 'change and improvement' in terms of a sufficient consensus of common purposes. As outlined such an approach requires a systemic alignment of the distinct if ultimately convergent axes of human knowledge-building.

The framework above is also useful for recognizing how various aspects of 'operational management' are integrally subordinate to not only what Knowles's calls 'lead-

Internal Axes------------System perspective------------External axes

Stakeholder consensus/ communication channels	Knowledge management *(human resource optimization)*	Science and technology innovations *(process and products of human knowledge)*	Adaptation to **changing economic vs. natural environments**
Develop sufficient convergence of perspectives, interests & general consensus in order to provide a foundation for a common commitment and purpose to an achievable outcome.	Encouraging, supporting and harnessing tacit knowledge of sector/ industry-focused stakeholders towards improved performance for overall or 'systemic change and improvement	To design and develop new solutions or adapt existing knowledge to new challenges and different contexts (technology as extension of body)	What changes or crises in society and nature represent an obstacles or challenge to be addressed to maintain or restore sustainability, viability and equitable sharing of resources
MACRO—in relation to distinct govt., commercial, civil society and academic perspectives/ partnerships	Linking small level problems and solutions to addressing larger problems and also developing solutions	Science as accumulated vs. applied social knowledge	Natural environment (dynamic homeostasis)
MACRO—internal to particular organization, industry, nation or even global level organization	Leadership/management develop a repertoire of micro interventions to tackle anticipated obstacles	Technology as extension of mind-body through tools, machines, and cultural-virtual networks	Vs. Changing social, economic and cultural contexts (growth)
Intergrity, communication, reciprocation	*Strategic planning/ systemic problem-solving*	*Experimentation/applied problem-solving*	*Observation*
Consensus-building (convergences despite us vs. them divergences)	Capacity-building (ecological vs. hierarchical optimization of human resources)	Applied knowledge-building-from data/ info to experience/ understanding	Sustainability-building (ecological optimization of natural resources)

Progress/profit//competition Vs. Sustainability/consensus/quality

Figure 4 *An enneagrammatic framework linking interdependent elements of integrated organizational/sector/industry problem-solving for sustainable policy development and optimal solutions. Adapted from Richards & Padfield (in press).*

ership strategies' but what we also would characterise as *sustainable policy problem-solving strategies*. Finally, this may also be discussed in terms of how the enneagrammatic structure of complex, systemic problem-solving also provides an 'emergent and convergent' systems framework for appropriating that well-known organizational epitome of 'integration, optimization and sustainability' the SWOT (strengths, weaknesses, opportunities and threats) analysis model. This is typically applied as an *us vs. them* (also *win-lose*) model of competition with strategic reference to some real or imagined adversary or opposition who/which might be defeated or overcome. However from an educational or other convergent win-win perspective (e.g., the community or global 'public good'), SWOT is ultimately much more useful as a framework to encourage every individual, group, and organization to realize its potential in an emergent and convergent way. Such an appropriation takes a more universal perspective of how people's 'strengths and weaknesses' ultimately tend to balance out. In this emergent and convergent way also, a much more powerful foundation can be achieved for harnessing the human potential, encouraging reciprocal dialogue, taking up new opportunities, and overcoming any perceived global as well as local 'threats'—and on this basis achieving also a common, deep-level, and 'enneagrammatic' future outcome or destiny.

CONCLUSION

In this paper we have attempted to outline how we think that our various efforts to construct an emergent and convergent model of effective problem-solving to tackle the endless 'wicked problems' facing the world are not only consistent with (a) the organizing principles of cutting-edge convergent science, but also (b) the related generic structures of ancient knowledge systems—as discussed, this is epitomized so powerfully by the enneagram model of process as outlined in the work of Bennett, Knowles, and others. The current international crises of sustainability are perhaps symptomatic of the inevitable failure of a 'non-ecological' modern world view. As Mandlebrot's fractal-based anticipation of the Global Financial Crisis outlined (Mandlebrot & Hudson, 2005), ultimately the behavior of financial markets as well as nature itself is fundamentally based on inevitably volatile and intrinsically problematic forces of change and not fixed structures of stability represented in mean averages. In other words it was not until the 'limits of growth' were globally obvious that a sufficient critical mass of awareness was perhaps possible for future transformation. But out of the discouraging ruins, confusing chaos, and complex dilemmas of such crises emergent new orders of knowledge ecology might be built based on principles consistent with ancient universal wisdom as well as nature itself. The enneagram thus remains an exemplary tool to guide this process of constructive transformation—that is, to guide integrated, optimal and sustainable problem-solving at both the macro and micro level.

REFERENCES

Aczel, A. (2000). *The Mystery of the Aleph: Mathematics, the Kabbalah, and the Search for Infinity*, ISBN 9780743422994.

Barrett, R. (2006). *Building a Values-Driven Organization: A Whole System Approach to Cultural Transformation*, ISBN 9780750679749.

Beke, G. (2008). "Gurdjieff and Greek esoteric thought," in *G.I. Gurdjieff, Armenian Roots, Global Branches*, ISBN 9781443800198.

Bennett, J. (1983). *Enneagram Studies*, ISBN 9780877285441.

Diamond, J. (2005). *Collapse: How Societies Choose to Fail or Succeed*, ISBN 9780143117001.

Fricke, M. (2009). "The knowledge pyramid: A critique of the DIKW hierarchy," *Journal of Information Science*, ISSN 0165-5515, 35(2): 131-142.

Gurdjieff, G. (c.1935). "The enneagram: A lecture," http://www.endlesssearch.co.uk/philo_enneagramtalk.htm.

Iamblichus (1988). *The Theology of Arithmetic: On the Mystical, Mathematical and Cosmological Symbolism of the First Ten Numbers*, ISBN 9780933999718.

Kahnemann, D. (2011). *Thinking, Fast and Slow*, ISBN 9780374275631.

Klein, J. (2004). "Interdisciplinary and complexity: An evolving partnership," *Emergence: Complexity & Organization*, ISSN 1521-3250, 6(1-2): 2-10.

Knowles, R. (2003). *The Leadership Dance*, ISBN 9780972120401.

Knowles, R. (2006). "The process enneagram, Parts 1-2," *Perspectives*, 20(1-2).

Kolko, J. (2012). *Wicked Problems: Problems Worth Solving*, ISBN 9780615593159.

Livio, M. (2002). *The Golden Ratio*, ISBN 9780767908160.

Mandlebrot, B. and Hudson, R. (2005). *The (Mis)Behavior of Markets*, ISBN 9780465043576.

Mitleton-Kelly, E. (2003). *Complex Systems and Evolutionary Perspectives on Organizations*, ISBN 9780080439570.

Pledge, K. (1983). "Structure process in scientific experiments," in J.G. Bennett (ed.), *Enneagram Studies*, ISBN 9780877285441, pp. 84-122.

Richards, C. (2010). "Socrates and 21St Century Knowledge-building," in A. Husain (ed.), *Psychological Explorations in Human Spirituality*, ISBN 9788182203167.

Richards, C. (2011a). "Global knowledge convergence: What, how and why the West has much to learn from 'the rest'," paper presented to the ELLTA Conference, Penang.

Richards, C. (2011b). "21st century knowledge building: The potentially crucial role of the humanities in the new university," *Kemanusiaan: Asian Journal of Humanities*, ISSN 1394-9330, 18(2): 19-41.

Richards, C. (2012a). "Sustainable policy making and implementation: Towards a new paradigm for a changing world," *Development Review*, ISSN 1607-8373, 21: 13-31.

Richards, C. (2012b). "Using a design research approach to investigate the knowledge-building implications of online social networking and other Web 2.0 technologies in higher education contexts," in N. Alias and S. Hashim (eds.), *Instructional Technology Research, Design and Development: Lessons from the Field*, ISBN 9781613501986, pp. 117-140.

Richards, C. (2012c). "Policy studies as framework for the renewed role of ethics in science and technology," *Philippiniana Sacra*, ISSN 0115-9577, 46(140): 409-442.

Richards, C. (2012d). "The most important new literacy? Overcoming seemingly impossible obstacles to make 'education for all' and related UNESCO goals and policies a reality in the 21st Century," *East West Journal of Business and Social Studies*, ISSN 2074-5443, 2: 123-158.

Richards, C. and Padfield, R. (in press). "Water as an exemplary focus of sustainable policy development: A Malaysian case study," *Journal of Environmental Policy and Planning*, ISSN 1523-908X,

Ricoeur, P. (1992). *Oneself as Another*, ISBN 9780226713298.

Rist, G. (2008). *The History of Development: From Western Origins to Global Faith*, ISBN 9781848131897 (2009).

Rycroft, D. and Kash R. (2004). *The Complexity Challenge: Technological Innovation for the 21st Century*, ISBN 9781855676114.

Schön, D. (1987). *Educating the Reflective Practitioner*, ISBN 9781555422202.

Schön, D. and Rein, M. (1994). *Frame Reflection: Towards the Resolution of Intractable Policy Controversies*, ISBN 9780465025121.

Scott, G. (ed.) (2002). *Does Socrates Have a Method? Rethinking the Elenchus*, ISBN 9780271021737.

West, J. (1993). *Serpent in the Sky: The High Wisdom of Ancient Egypt*, ISBN 9780835606912.

Chapter 3

2008 MECS SUMMIT: A WORKSHOP ON COMPLEX SITUATIONS

Beverly G. McCarter

This paper examines a unique and modified form of the Process Enneagram© that was designed to tackle the issue of examining underlying principles of Wicked Problems in an effort to see them from different perspectives and, hopefully, to gain greater insight into practical methodologies that might guide our efforts to deal with them. This modified form of the Process Enneagram© was specifically designed and used at the invitation only 2 day Managing and Engineering Complex Situations workshop conducted in 2008 and hosted at the MITRE location in McLean, VA just outside Washington, DC.

MECS FORUM

The Managing and Engineering Complex Situations (MECS) forum was created to bring a diversity of practitioners from academia, industry, and government together to advance the way problems are solved. Focused on change and innovation, it explored new methods for understanding and coping with complexity.

In October 2007 the first MECS conference was held. It focused on the complexity of wicked problems and its influence on problems today. The term "wicked problems" was originally defined by Horst Rittel and is defined as problems "...for which each attempt to create a solution changes the understanding of the problem" (CogNexus Institute, http://cognexus.org/id42.htm).

Because of the interconnected and ever evolving nature of the variables involved, wicked problems cannot be solved in traditional linear methods. Wicked problems always occur in the context of human systems, reflecting the diversity of perspectives, paradigms, and views of reality of the individuals involved (CogNexus Institute, http://cognexus.org/id42.htm). It is this social complexity of problems in today's hyper-connected world that overwhelms problem solving in organizational structures.

"The concept of "wicked problems" in design was originally proposed by H. J. Rittel and M. M. Webber (1984) in the context of social planning. They pointed out that in solving a wicked problem, the solution of one aspect may reveal another, more complex problem. Rittel and Webber suggested that the following rules define the form of a wicked problem:

1. There is no definitive formulation of a wicked problem.
2. Wicked problems have no stopping rule.
3. Solutions to wicked problems are not true-or-false, but good-or-bad.
4. There is no immediate and no ultimate test of a solution to a wicked problem.
5. Every solution to a wicked problem is a "one-shot operation"; because there is no opportunity to learn by trial-and-error, every attempt counts significantly.
6. Wicked problems do not have an enumerable (or an exhaustively describable) set of potential solutions, nor is there a well-described set of permissible operations that may be incorporated into the plan.
7. Every wicked problem is essentially unique.
8. Every wicked problem can be considered to be a symptom of another problem.

9. The existence of a discrepancy in representing a wicked problem can be explained in numerous ways. The choice of explanation determines the nature of the problem's resolution.

10. The planner (designer) has no right to be wrong" (Computational Complexity and Problem Hierarchy, http://cs.wallawalla.edu/~aabyan/Theory/complexity.html).

In July 2008 the forum continued exploring this issue with an invitation only 2 day workshop hosted by MITRE at its McLean, VA complex. The General Council for the workshop included Dr. Andres Sousa-Poza, Samuel F. Kovacic, Dr. Adrian Gheorghe, Beverly Gay McCarter, Dr. Brian White, and Dr. Chuck Keating. The MECS Forum Committee included Beverly Gay McCarter, Dr. Lowell Christy, and Dr. Dennis Buede.

The 2008 workshop's theme was "Identifying the Paradigm Shift". Its purpose was to engage a theoretical discussion of complexity that would lead to guiding principles for handling complex situations. This workshop explored effecting a paradigm shift through conversation of how we understand complex problems by examining what we do not know. The diversity of the individuals invited to attend this workshop, and their respective viewpoints, was an important element in helping to facilitate this paradigm shift. It was hoped that the conversation amongst the participants would help untangle the knots of understanding wicked problems through fostering a common understanding of the issues involved: creating a paradigm shift in our perceptions of how to solve complex problems.

STRUCTURE OF THE WORKSHOP

The 2 day workshop began with introductory remarks that set the stage for what defining the purpose and hopes of the workshop. Participants were then divided into 4 groups for discussion purposes over the next 2 days. It was decided that Richard Knowles's Process Enneagram© would be used to facilitate the discussions since its design was intended specifically to help individuals untangle the inherent complexity in wicked problems, move individuals through transformational change by helping to facilitate rational decisions, build relationships, and create emotional energy around the decisions made. The group facilitators for the workshop included:

- Group 1—Dr. Richard Knowles, Richard N. Knowles and Associates; developer of the Process Enneagram©;

- Group 2—Beverly Gay McCarter, Human Mosaic Systems, LLC; facilitator, certified in the Process Enneagram©;

- Group 3—Bill Miller, WDM Systems;
- Group 4—Dr. Brian White, MITRE.

At the end of the first day, the groups presented updates to the whole body of participants as to their discussions. It was hoped that this diversity of ideas and perspectives would help add greater insights to the final group discussions the following day. At the end of the final day the 4 groups presented the final results of their discussions. Andres Sousa-Poza then led a discussion of next steps to take to move the groups' suggested actions forward.

Opening Remarks

Below (with permissions granted) are the opening remarks for the conference that set the stage for the subsequent discussions.

Dr. Andres Souza-Poza, Old Dominion University

Welcome. I want to thank MITRE for having the facilities and the committee that put this together. I want to look at what we are going to be doing. Everyone in this room has been invited because of their level of expertise. Everyone has experience in dealing with these types of problems. We have a wide range of issues.

The most important aspect of MECS is its participants. MECS is a self defining emerging construct. What comes out of this workshop will define the purpose of MECS and who we are. I am a mechanical engineer. I enjoy things that are predefined and I am anxious when they are not. The intent is to synthesize a product out of this workshop. We have an excellent group of people here to help start working these problems. It's an emergent process. Thank you for coming.

(Permission to reprint here given by Dr. Souza-Poza, 2012)

Dr. Lowell Christy, Cultural Strategies Institute

Thank you. I'm with the Cultural Strategies Institute. We want to understand the complexities of "human in the loop". I want to answer why we are here. Every ten years since WWII, we have had an ominous "c" word. Cybernetics. Club of Rome. Catastrophe theory. Now, complexity. Underlying each one is a constant theme—our rational norms are not adequate to deal with phenomena.

We all understand when bad ideas kill real people. In 1947, a group of people decided after the war we needed a better way to make human decisions. The Macy Foundation funded this effort. The Office of Naval Research gathered a group in Chicago in 1959 to find a better way to win the Cold War. What's happening now is that we can see the bald spot on a terrorist's head, but we can't see what's going on inside it. We were constrained in Vietnam, and went for "shock and awe" in Baghdad. We didn't allow for differences in cultures.

We need to understand how technologies interact with the human element. Human systems need to be understood better, and that's part of why we are here today. We are the inheritors of thinking that has brought changes to the world. We need to think about the notions of change. Complicated change is putting humans on the moon. Complexity is teaching Johnny to read.

1622, 1776, 1789. Three dates created the world we have here. In 1622 Francis Bacon wrote "Nova....", on the interaction among educated people. 1776 was the American revolution, which was "doomed to fail". Adam Smith wrote of the virtuous nation. In 1789 we had the political interaction that came out of our Constitution and the French revolution. How do we work with others to have a more purposeful world, a larger ecology?

Our challenge is a "many minds" problem, how to interact virtuously instead of viciously.

(Permission to reprint here given by Dr. Lowell Christy, 2012)

Emergence: Living on the Edge of Chaos—Dr. Richard N. Knowles, Richard N. Knowles and Associates

I came to this work on complexity after working in research at DuPont. Why do people get tangled up in groups, when that is not their intention? What happens when there is a crisis at a chemical plant? People were pleased about how people worked together during the crisis, but we can't burn a plant down every six months in order to feel good.

When I went to the first chaos conference I attended, I started to meet people who were talking about chaos and complexity. I want to engage you in a conversation here. This stuff is fuzzy and we don't know where we will land with it. Please bring your backgrounds to this effort. We need to come together and listen to each other, suspend our judgment, talk together, be fully present, and speak your truth. That's not a license for personal attacks. Let's come together out of humility.

I'm not in organizational development, I just try to figure out how to do things better. People are self organizing all the time. What are they going to engage in and self-organize around? What's important to them. As leaders we have a simple choice: we want to engage this process in a way that's purposeful, or we can impose our will. For many years, most have imposed their will. Maybe that's not always the answer.

Imposing our will imposes order on a chaotic situation. We get consistent patterns of behavior. As we work with organizations, a big challenge is to open the conversation but not have it blow up on us. We move toward emergence in various bifurcations. We're not asking for universal buy-in. We start with a challenge. Do we want to impose answers or open it up and see what emerges? Whatever happens at the conference will be something we have created together.

Self organization happens all the time; it's pervasive. Leadership is temporal and occurs in the moment. We co-create our future together. There are many ways to have the conversation and engage together. We want to use the process enneagram. The enneagram is also used in personality presentations, but we are looking at process. What is the process of work? The first thing we talk about is identity. Who are we as individuals, and how did we get here?

Next is relationships—can we become interdependent? Third is information; how do we create and share it? These are the three basic elements of self-organization.

Then we begin to talk about intention; what are we trying to do? Tensions and issues: what causes concern? What are the principles and standards of behavior to which we're willing to subscribe? As we work on these issues, we create value. All the ideas on the Process Enneagram are connected; some connections are more important than others.

We will start with identity and intention; who are we and what do we want to have happen over the next two days. Standards will be developed to take us there. What barriers do we have to overcome? It may be languages, or technologies, or discomfort over ambiguities. Some of these tensions are unresolvable. [sic] We can live with ambiguity if we are listening and learning.

How do we organize ourselves around MECS? Can it become a living organism going into the future, or will it die here? What do we really mean about the word complexity? At the end, we need time to reflect on what we've done. We don't lack technical skills, we lack the ability and time to sit down and work these things out. Let's try this process in our time here together. I have used this process when several chemical plants talked to a community of 300,000 in West Virginia about what procedures would be taken in the event of emergency. It worked very well, and let's try it here together.

(Permission to reprint here given by Dr. Richard N. Knowles, 2012)

USING THE PROCESS ENNEAGRAM© FOR MECS

Figure 1 is an example of a simple Process Enneagram© map that includes the 9 interconnected variables that are simultaneously at play in dynamic complex systems and what questions are explored with each variable to help the group move forward in facilitating their collaborative efforts. The Process Enneagram©, in wide use around the world since the 1990s, is a successful process developed by Dr. Richard (Dick) N. Knowles for understanding simultaneous complex dynamics at play in human systems, particularly those found in organizations. It enables these dynamics to be mapped, identifying the current state of an organization, and helping individuals collaboratively develop action plans to move forward while building relationships and creating emotional energy and commitment to the plans developed.

The process allows individuals and organizations to be flexible and adaptable in their thinking, analyzing complex problems from a variety of perspectives. In the problem analysis, it allows individuals to see what is not working in their organization and why. Individuals are able to have their individual perspectives heard as the group collaborates to create plans to move the organization forward to achieve their goals. The Process Enneagram© allows individuals and organizations to look at the whole picture of the complex dynamics affecting an organization and to examine the key individual dynamics without being overwhelmed by the confusion of the simultaneous dynamic interactions. As a result, it allows an understanding of complex processes and creates a paradigm shift in the way individuals approach problem solving in complex environments.

The strength of the Process Enneagram© is the facilitated guided conversation. It connects people with diverse perspectives and roles, allowing each to be heard and to have a voice in the expanding discussion. It incorporates basic counseling psychol-

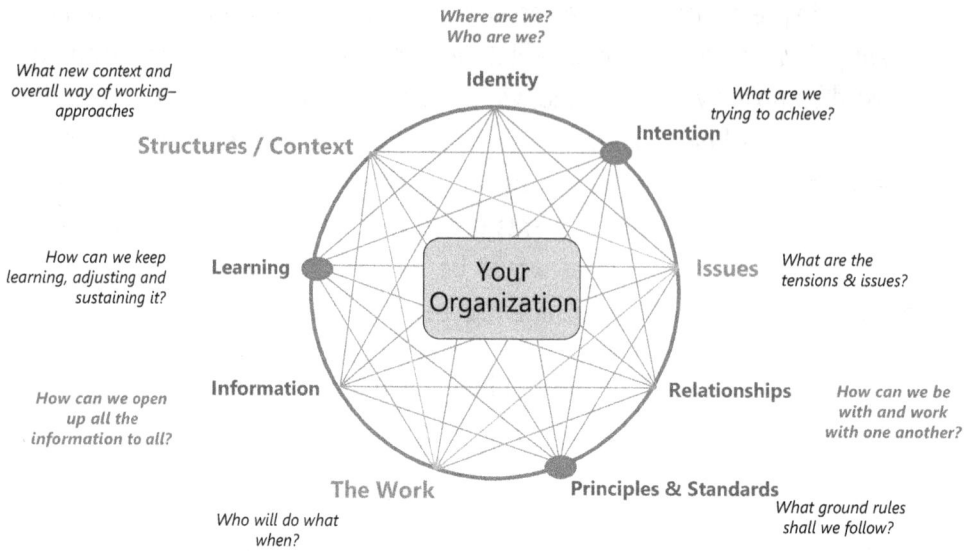

Figure 1 *Example of a Simple Process Enneagram©.*
Source: R.N. Knowles, "The Leadership Dance".

ogy principles and methodology: understanding oneself and others, facilitating the building of relationships through sharing information, and understanding one's role in helping the group achieve its goals. (http://www.centerforselforganizingleadership. com/pages/processenneagram.shtml, and http://www.centerforselforganizingleadership.com/pages/articles.php?Why%20the%20Process%20Enneagram%20works!)

The process enables emergence, new ideas and solutions arise from this self-organizing process. In today's complex hyper-connected world, the conversation among people has a greater influence than ever before in affecting the ability of organizations to navigate an ever changing complex environment.

As shown in Figure 2, the Process Enneagram© is mapped in a circle to demonstrate that complex systems are not linear. They involve dynamics that are happening simultaneously. The circle is referred to as the "Bowl" which helps to give the self-organizing group some structure to explore the complex dynamics by examining those variables in the context of an overarching question, which can be very specific or very abstract that is posed. That question becomes the Bowl and the nine dimensions of the map are explored in a specific order to answer that question. Without the structure or purpose of the question it becomes very difficult to begin to understand a complex system. The Bowl gives the group a sense of shared identity and purpose or intention. They define who they are and what they are trying to do. This definition of who the group is helps to create interdependence among the group members which,

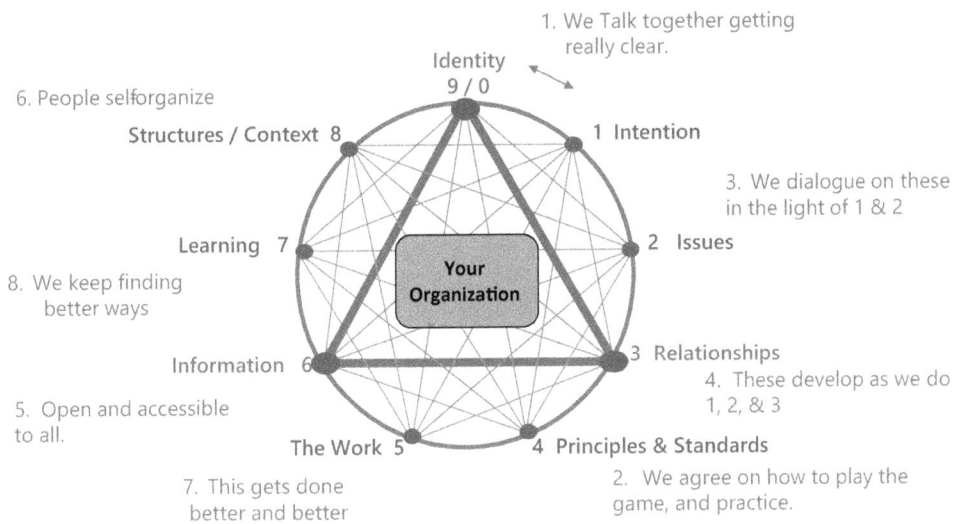

Figure 2 *Process for Self-Organization.*
Source: R.N. Knowles, "The Leadership Dance".

in turn, helps to improve their relationships within the group. Establishing this Overarching Question is the first step the Process Enneagram© session.

Figure 2 illustrates the heart of the Living System, the core of how human systems operate: Identity, Relationships, and Information. It also shows how the Living System interrelates with the remaining 6 variables in order to effect a shift in the way people understand a complex problem and collaborate in a self-organizing manner to create solutions to those problems.

Figure 3, The Domains of Self-Organization, highlights the heart of the process, the Living System: Identity, Relationships, and Information. It is the basis of human systems upon which conversation and collaboration depend. It is what drives Emergence in Complex Systems.

Identity involves a sense of one's self as well as others. Here, it also looks at the group's shared history together. Who are they in light of their roles within the organization or relationship? Identity is labeled both 0 (the starting point) and 9 (the ending point) because the process is a cyclical one where learning from the environment is built in and iterative. As a complex system changes, so, too, will the 9 variables within that system change. As a result, the re-evaluation of those variables and the problem solution is always occurring. Individuals must be flexible to be successful in complex systems, and so, too, must organizations. Relationships (point 3) examine what the relationships are like within an organization. Are they healthy sharing information,

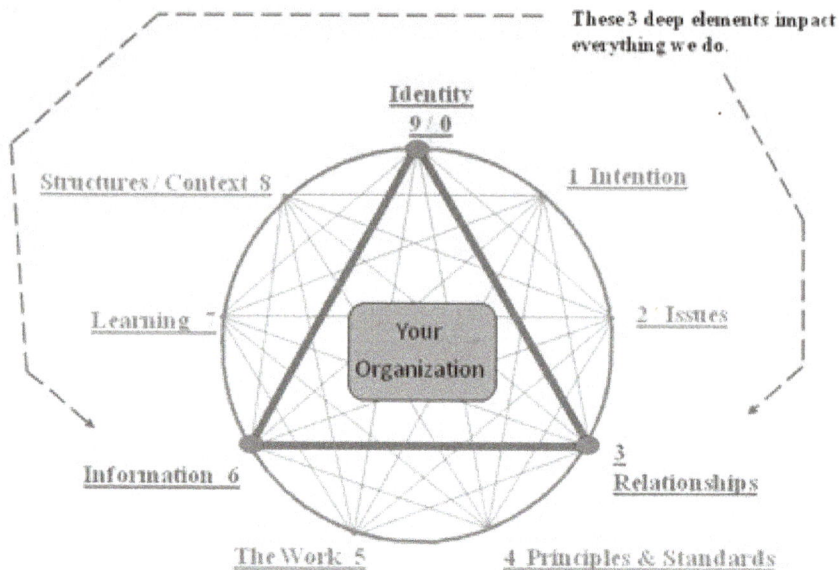

Figure 3 *The Domains of Self-Organization*

getting work done, and networked efficiently within the system? Through the Process Enneagram© these relationships grow and develop. And finally, point 6, Information. This variable examines whether or not individuals within the organization are able to know what is happening within the organization. Are they isolated and stove-piped, or is information freely flowing? How do the individuals within the organization handle information? These 3 variables make up the Living System that is the core of complex adaptive systems: human interaction. It is The Conversation.

THE PROCESS ENNEAGRAM© MAPS FOR THE MECS WORKSHOP

For the 2008 MECS Workshop we used the Process Enneagram© to help facilitate a group discussion that would help formulate universal rules that define complex systems and how we might be able to better handle Wicked Problems. The invited participants of the conference are experts in complexity from a wide range of fields including Art, Psychology, Mathematics, Physics, Engineering, Systems Engineering, Biochemistry, and Philosophy, to name a few. We sought to have the conference participants share what they knew from their own unique perspectives.

Each unique perspective or way of looking at how they came to understand complexity, in part, was based on an immersion process; based on experience. Each of us gains an understanding of complexity through an experiential process that makes this understanding seem intuitive. Through our lifetime of experiences, we pick up bits and pieces of data along the way. Those bits and pieces are subconsciously connected at a higher level of brain processing that is often referred to as intuition. This intuitive way of understanding (a "gut feeling) is not readily articulated in words. One just "knows" it is true. Also, each person has a unique view of reality and understanding of complexity based on those experiences in life, as well as their physiological and psychological makeup.

The task of the MECS conference was to enable this diverse group of incredibly creative thinkers to be able to articulate what they intuitively knew. They needed to be able to articulate it in a manner that others could understand despite semantic differences. What was the universal language to verbally express what they all uniquely and intuitively knew? What was the underlying universal pattern of understanding complexity that they all shared?

Richard Knowles and Beverly Gay McCarter created a modified Process Enneagram© (PE) map to help facilitate this challenging task. They modified the traditional PE to better suit this unique conversation.

The groups of diverse individuals gathered for this discussion are all experts in the field. Our goal was to facilitate this group of incredibly creative thinkers to articulate what they intuitively knew in a way all could understand. The ultimate goal was to try to identify the underlying theories that would enable us to handle complex problems in a more practical manner. Is there an unarticulated methodology that underlies the theory that could better guide our efforts to deal with wicked problems? And if there are underlying rules or laws to dealing with complexity, how should we be dealing with them?

In modifying the PE for this session, we began by looking at the following questions:

- How did we get to where we are?
- How do we recognize where we are?
- How do we verbalize where we are?
- What are the implications of this environment?
- Are there new ways of dealing with it?

With these in mind the following 5 topics were incorporated into the traditional PE to help begin the conversation to create a paradigm shift in the way we looked at Wicked Problems:

THE OVERARCHING QUESTION: What are the conditions of working in complex situations or having to address wicked problems which lead to the best chances for success?

- (Traditional PE connection: This is the "Bowl" discussed earlier that helps define what question the group wants to address and helps develop cohesion within the group)

The Problem

1. What are the derived conditions of working in complex situations or having to address wicked problems? (What do we experience? What are the conditions we face? What is the reality?)

   (Traditional PE connection: This corresponded to the Identity 9/0 variable of the PE in that they looked at their personal experiences and how they came to intuitively understand complexity.)

The Approaches

2. What in our present paradigms inhibits us from effectively dealing with the problem? (What is not currently working in the way we view complexity and try to deal with it? What are our current organizational structures?)

3. What should we be doing to better deal with the complex situations? (What do we need to be doing differently in approaching the problem?)

   (Traditional PE connection: This corresponded to the Issues 2 variable of examining the tensions and issues at play within the organization.)

The Principles

4. What principles can we derive from the ways that we propose to deal with complex situations? (What are the universal patterns that are derived from different perspectives?)

   (Traditional PE connection: This corresponded to the Principles and Standards 4 and The Work 5 variables where you identify the ground rules to follow in how individuals work together to move the organization forward and then decide what tasks need to be done to move accomplish the stated goals. The Principles and Standards at point 4 are the key for helping us to live in the ambiguity and tensions at point 2.)

5. What implications do these principles have on existing methods and approaches? (What is impacted negatively by the current methods and approaches? What needs to change? How do we change? What do we need to do differently?)

6. How can the generated principles and rules be institutionalized in organizations and individuals? (How do we help bring about effective change to better handle complexity? How do we need to impact our organizational structures?)

   (Traditional PE connection: This corresponded to the Learning and Potential 7 and Information 6 variables where the organization examines how they share information to better enable the collaborative process and what processes are in place to enable the group to continue to learn and adjust to meet the demands of a changing complex environment.)

Beginning with the PE variable of Identity, the conversation unfolded initially with the use of Story Telling techniques which enabled the participants to articulate how they came to understand complexity. What was their unique and intuitive experience? What patterns did they sense that were common to their understanding and experiences of complexity?

Story telling is a good method for tapping into this understanding. It enables participants to articulate the way they came to understand complexity through their life experiences. By telling the stories, we hoped to be able to by-pass arguments about semantics because our individual experiences are unique and our own; we also hoped to find a common ground for understanding in this diverse group of participants; and we hoped to tease out common patterns or principles of complexity common in many of the individual stories.

Then, through guided facilitation, the 4 groups began 2 days of engaged and, at times, spirited discussions.

Results

A summary of the collective group responses for the questions posed at the 2008 MECS Workshop are as follows:

The Problem

1. What are the derived conditions of working in complex situations or having to address wicked problems? (What do we experience? What are the conditions we face? What is the reality?)

    Collective response:

    -   Because of emergence in complex systems, reductionism won't work for complex situations, though it may work for systems within the complex situation.
        -   Engineers used to solve problems; different processes and methodologies (traditional, collaborative, etc).
        -   Solutions today are broader than just engineering solutions.
        -   The term systems implies predictability which is not possible in complex situations.
        -   Complex situations involve an interactive environment/variables that results in emergence.
    -   Due to the diversity of individual perspectives and culture found in complex situations, individuals experience different aspects of engagement.
        -   The level of engagement is based on one's level of understanding.
        -   Nontraditional methods are needed to deal with complex situations.
        -   There is a cultural impact on trans-global complex systems.
        -   Assumptions need to be questioned.
        -   Ever changing complex situations require flexibility and adaptability.

The Approaches

2. What in our present paradigms inhibits us from effectively dealing with the problem? (What is not currently working in the way we view complexity and try to deal with it? What are our current organizational structures?)

    Collective response:

    -   The dominant discourse is that systems thinking that it is the only/final solution.
        -   Systems thinking is not the holy grail;
        -   "Close enough" worked in the past, but not anymore with complex situations and human interactions.

- A complex situation consists of an irreducible dynamic human component.
  - Human interaction is a non-systemic dynamic : it doesn't work with traditional systems methodology. Due to globalization human interactions and their unpredictability are now more dominant and drive complex systems.
  - Diverse perspectives can lead to conflict as individuals reject perspectives different than their own.
    - Inflexible attitudes.
    - Need to acknowledge different perspectives.
    - Need to understand a person's underlying philosophy in order to effect a change.
    - Advocate by example, be the change you wish to see, then enter into a dialog
  - People are not rational, and are not predictable.
- The way that we justify what we believe is through knowledge; Rationality/ scientific method is not going to help us understand complex situations. If we accept the fact that people are unknowable we will always dabble with positivist methods of truth.
  - Scientific research assumes there is order in the universe; but, the only thing it will reveal to us is order, as a result.
    - Assumptions get lost and become unavailable
  - Our methods for finding truth are limiting us now. Based on what is known. But what is the unknown?
    - The methods we know inhibit us in exploring what we don't know.
    - We have lost our ability to critically think. We make decisions without thinking.
    - Methods we have for acquiring knowledge are insufficient today.
    - We are limited in our access to information.

3. What should we be doing to better deal with the complex situations? (What do we need to be doing differently in approaching the problem?)

Collective response:

- We need new ways of working together to address the interface between complicated and complex situations.

- We need to translate our ideas about how people work together on complex situations to the way systems work together.
- We need to develop a multidisciplinary environment to understand complex situations.
- Embracing chaos in certain situations can encourage innovation.
- We must engage individuals in conversation—understand the inherent dynamics.
    - Need to strive for effective communication between individuals.
    - Individuals need to participate in the solutions—take advantage of the wisdom in diverse perspectives for solutions.
    - Our perspective colors our engagement.
    - Need to be able to appreciate differences in culture, experience, background of participants.
    - Individuals must be able to morph, be flexible, be able to adapt, flow with the moment, to test options, and to meet changing needs.
    - Groups need to strive for a shared reality, acknowledging that perception influences our personal reality.
    - Individual diversity in expertise and perceptions enable groups to achieve richer and more innovative outputs.
- There are changing notions of success and measuring success for complex situations. Human dynamics are not readily quantifiable.
    - Develop case examples, capture success / unsuccessful stories that illustrate key aspects of the complex system challenge with unique perspectives.

The Principles

4. What principles can we derive from the ways that we propose to deal with complex situations? (What are the universal patterns that are derived from different perspectives?)

Collective response:

- Complexity theory might change our understanding of stability... it may be a fluctuation around an attractor, it's not fixed.
    - There are shifting islands of stability within a sea of instability in complex situations.

- Orderliness is defined by a chaotic dynamic, a different sense of stability within a shell.
- Can't talk about a single point of stability; stability is within the shell; the shell is stable within its structure, but it's not simple.
- Individuals must learn to live within a stability of the "shell".

- Do we need a new institution to champion learning about complex systems?
- Responsiveness aspect of humans responding based on individual viewpoints is important to consider.
- Metaphors can be powerful for learning in complex dynamics, as well as for teaching about diversity (a minimal requirement for creativity, growth, intentionality, and transformation):
  - Metaphors and storytelling help us to understand complex situations;
  - Need to encourage storytelling to help explain complexity;
  - Storytelling is the anecdotal sharing of information.

- The interaction of human and machine creates new and emergent patterns.
- Interactions of individuals in groups can result in unpredictable emergent group behavior.
  - Schismogenic systems.
  - The interaction or conversation among individuals drives complex situations through emergence.
  - Diversity of perceptions facilitates emergence.

- Characteristics of complex system includes: high throughput, a "living" system, energy/information flow in and out, richness/diversity, and interactions are key.
  - Complex situations include dynamic self assembly and self repairing.
  - Complex situations have no beginning or end; they are evolutionary and continuous and do not follow traditional life cycles.

The Way Forward

5. What implications do these principles have on existing methods and approaches? (What is impacted negatively by the current methods and approaches? What needs to change? How do we change? What do we need to do differently?)

Collective response:

- Complexity includes emerging known and unpredictable systemic characteristics:
    - Individuals and organizations must be able to morph, be flexible, be able to adapt and flow with the moment, to test options, and to meet changing needs.
- A common vocabulary is needed, a commonality of perspective to communicate effectively:
    - Definitions, lexicons, frequently derail solutions;
    - Differing vocabularies provide richness to the conversation;
    - From Second Order Cybernetics and lateral thinking—our understanding of what is science is changing:
        - "The epistemologies of realism, constructivism, and pragmatism can be viewed as emphasizing different combinations of world, description and observer. Realism emphasizes world and description. Constructivism emphasizes observer and description. Pragmatism emphasizes observer and world. These three epistemologies are similar to three stages in the development of cybernetics—engineering cybernetics, biological cybernetics, and social cybernetics. Viewing the three epistemologies as emphasizing different facets of a triangle clarifies the relationships among the epistemologies and creates an opportunity for unifying them. Advocates of each point of view tend to direct a conversation toward the issues of greatest interest in that epistemology." Stuart Umpleby, Unifying Epistemologies by Combining World, Description and Observer, (http://www.gwu.edu/~umpleby/recent_papers/2007_ASC_Unifying_Epistemologies.pdf)
        - Multiple views of knowledge or perspectives enhance a larger view or understanding of complex situations. Individual perspectives impact our understanding of reality.
        - In order to understand complex situations, views of the world, descriptions, and the Observer (the perspective of the person analyzing the data) must be included.
        - All three views have to be included in our understanding of complex systems.
        - How do we include all three views when the observer is in the system, outside the system, and on the border between them?

- 4 derivatives to pass through to understand complex situations: linear; recursive (non-linear, cellular automata); observer; social.
  - Three views need to be included to help understand complex situations better:
    - Engineering;
    - Perception, interpretation, biology;
    - Social purpose, reflectivity(feedback);
    - More information on this topic (http://ebookbrowse.com/2010-wmsci-science-2-ppt-d112851965).
  - Implications are that you have to
    - Mature observations to principles, then to action; but, this takes time.
    - However, once they are enacted, the situation has changed, yet again
- Loose, virtual and adaptive structures are appropriate.
- One structure does not fit all—combination of hierarchy and distributed dynamic network (hybrid).
- Social (and anti-social) networks are a key consideration of complex systems.
- Interactions between social and institutional networks for a given purpose define emerging patterns.
- Communication level and nature need to be appropriate to the purpose and pattern.
- How can social networking tools and techniques help?

6. How can the generated principles and rules be institutionalized in organizations and individuals? (How do we help bring about effective change to better handle complexity? How do we need to impact our organizational structures?)

Collective response:
- Need an open multidisciplinary forum where ideas can be shared:
  - Mixed forums;
  - Discussion groups;
  - Network with other groups—links with other groups:
    - Needed collaborative workspace to share information so the conversation can continue forward.
- Provide examples, case studies (metaphors and storytelling):
  - Learn from pitfalls.

- Develop relationships;
- Committees needed:
  - "Wicked problems"
    - Committee to make explicit the characteristics of wicked problems—the extreme example
  - Manifesto/ white paper of principles committee;
  - Image conceptual committee;
  - Distribution via network committee;
  - Network with other group committee (link with other groups);
  - Funding committee.
- Asking for cultural change:
  - How to change a culture in organizations:
    - Train in organization—change understanding;
    - Change diversity within organization;
    - Bottom-up change.
- Learn with training: (knowledge transfer for activity)
  - Is the knowledge transferable?
  - Need a product: then distribute it:
    - Lowell's documentary... book, journal, film, music
    - Loose intellectual property rights—free transfer
- Innovation occurs—network changes—diffusion of innovation.
  - Hubs/nodes: like Linux
- Make it free to spread change—
- Need a model:
  - Example
    - "Elegant Universe" model—website
    - "Earthrise" photo by Apollo astronauts—single image
      - Craig Hayenga—SFI use of artists to pictorially conceptualize complex concepts
  - Must be simple, short, to the point
  - Image by conceptual committee needs to be drawn up for MECS
  - Manifesto/ document of principles committee
  - Distribution via network committee

- ◦ Network with other group committee (link with other groups)
- ◦ Funding committee
- • Need to put stories/parable on a website-Craig Hayenga, Stuart Umpleby
  - ◦ (MECS name change—not acronym anymore)
  - ◦ How do we get new customers to hear us?
  - ◦ How to combat the bias against engineers and our own inherent biases?

KEY LEARNING FROM THIS USE OF A MODIFIED PROCESS ENNEAGRAM©

The PE is a very flexible process. Its key strength is its underlying structure that is based in counseling psychology principles and methods that facilitate individuals understanding their own cognitions and being willing to listen to others' viewpoints. The ability to "see", learn, and make decisions on new information is key for adaptability in ever changing environments. The basic 9 variables of the PE are flexible in their applications because they represent key universal aspects of complex human dynamics, particularly where problem solving is concerned. The heart of the process is called the Living System; identity, relationships, and information are key structures in understanding wicked problems. It doesn't matter what the individual experiences or perceptions may be. What matters is actually hearing what others say and using shared information to find a way to move forward.

As a result, using the PE to facilitate the conversation was a very good fit for our purpose for the MECS workshop in 2008.

The MECS workshop took a facilitative process for untangling complex human systems in organizational structures and applied it to the problem of individual perceptions of a very difficult and abstract problem. The reason it was able to do this was because it examines the inherent universal variables found in complex human systems and how our perceptions and theories of how the world works are impacted by these same variables. These key issues at the heart of complex human systems include:

1. Identify your perceptions, experiences, role and what you want to accomplish. This is the "I" part of the process;

2. What are the tensions and issues involved with what you are examining or trying to accomplish;

3. What ground rules need to be developed to facilitate accomplishing the stated goals and what work needs to be done to move the mission forward;

4. What isn't working and needs to be changed, and how do we facilitate that change?

The specific questions asked or problems examined using the PE can vary widely. What is similar is that they involve humans in the loop and endeavor to move complex human systems forward.

Much has changed since the MECS 2008 Workshop as more people come to recognize and now tackle the challenge of understanding complex systems. A global dialogue is taking place though it is still not completely networked.

The power of multi-user immersive virtual environments in helping individuals understand emergence in group dynamics is being explored, as well as the power of using storytelling and metaphor to understand complex dynamics. (See recent articles from IARPA, DARPA, and the Pentagon: IARPA Sirius Project http://www.iarpa.gov/rfi_sirius.html, DARPA STORyNET http://www.wired.com/beyond_the_beyond/2011/02/design-fiction-special-notice-darpa-sn-11-20-stories-neuroscience-and-experimental-technologies-storynet-analysis-and-decomposition-of-narratives-in-security-contexts/ , as well as the Pentagon and the use of theatre http://www.berkeleyrep.org/press/pr/1011/Berkeley_Rep_Great_Game_Pentagon.pdf)

The thoughts and perspectives gleaned from the 2008 workshop are already finding their way into the general understanding on the topic within the community at large. The conversation is well underway. And the use of the PE as a process to facilitate interactions and the understanding of oneself, others, and group dynamics combined with metaphor and storytelling to understand complex dynamics (especially in a kinetic immersive 3D environment which allows participants to become a part of the story and not merely an observer) can effect many of the cultural changes the workshop referred to in order to advance a greater understand of complex situations and how to navigate them.

The future is here and the conversation has started in earnest.

ACKNOWLEDGEMENTS

I would like to thank Dr. Richard N. Knowles, author of "The Leadership Dance, Pathways to Extraordinary Organizational Effectiveness", for introducing me to this insightful and powerful facilitative process for dealing with complex human systems (Wicked Problems). Its strong Counseling Psychology undercurrents helped me to instantly understand and appreciate its capability to facilitate transformational change in complex systems.

Chapter 4

THE PROCESS ENNEAGRAM©
A PRACTITIONER'S GUIDE TO ITS
USE AS A FACILITATIVE TOOL IN THE
CORPORATE ENVIRONMENT

Catherine Taylor

With today's dynamic work environments complex problems arise more frequently. Indeed the very notion of complexity underlies so much of modern discourse on organizational behavior. But ways of capturing the fundamental idea or lens of complexity theory and turning it into a useful and pragmatic way of solving such problems are few and far between. One very useful tool to tackling these issues is The Process Enneagram. This paper explores some applications of the Process Enneagram following years of practical experience in its use, as well as highlighting both its strengths, areas for caution and learnings gained from experience with this model.

INTRODUCTION

The Process Enneagram fulfills a number of purposes: a conceptual schema, a way of thinking about emergent systems, a methodology for analysis, and a tool to guide facilitation and dialogue, among others. It is this last use, as a process tool, that this paper will explore.

The Process Enneagram was developed and refined by Richard N. Knowles (2002) originating through his experience with DuPont in Belle, West Virginia starting in 1987. This article aims to highlight some of the advantages and learnings from using the Process Enneagram. It takes a deliberately pragmatic and reflective approach to its use in real workplace settings[1]. It also seeks to identify matters to which one should

---

1. For background context on the more conceptual aspects of the Process Enneagram the reader is re-

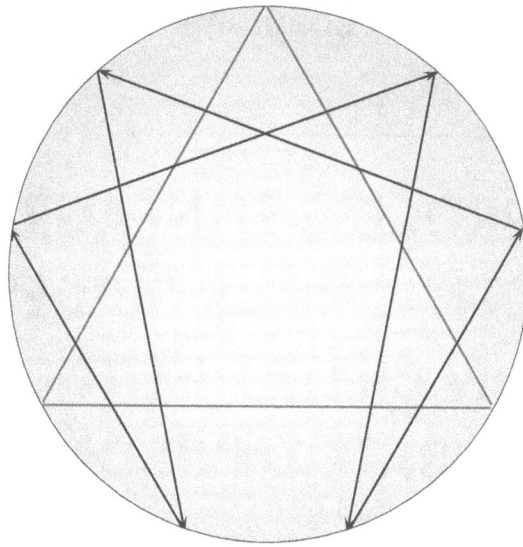

Figure 1 *Schematic of a Basic Enneagram.*

pay attention in order to ensure success, particularly as a group facilitative framework. And ... this paper also seeks to share some pitfalls and solutions I have had the fortune to come across which, once understood, allow the Process Enneagram to be used even more productively.

This paper arises from my experience as a management consultant and trainer, using the Process Enneagram with groups and individuals, as well as individually, for the past 10 years. I have used it professionally with individuals and groups ranging from 2 people to 40 people, although this in no way sets limits as to its application. It has been used in corporations from a variety of industries across many countries, not limited to but including Australia, New Zealand, the USA, Malaysia and the UK.

It is worth noting at this stage that the particular labels I use in the Process Enneagram framework are essentially the same as those used by Knowles, but with slight changes and additions. It is considered these labeling variations are insignificant to its effectiveness of use and to this paper. The particular elements upon which this paper is based are illustrated in Figure 1.

ferred to Knowles (2002), Dalmau & Tideman (2011).

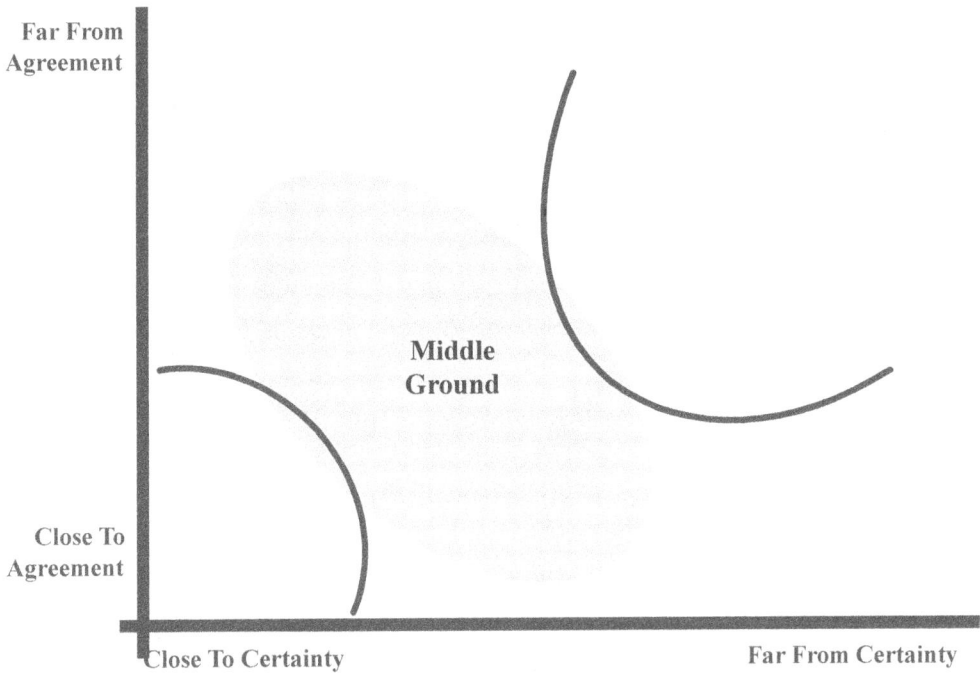

Figure 2 *Ways of Classifying Problems.*

CAUTION

This paper is written in the true belief that little if anything can be predicted in corporate life. Stacey (2012) points out that when people talk about a tool for management or leadership in organizations in the hope they will enhance something, the underlying, taken-for-granted assumption is that *"if you apply tool M you will get result X"*, i.e., a predictable outcome. Stacey (2012) goes on to point out that organizations do not take linear form and efficient causality does not apply (p. 53).

I consider the Process Enneagram to be an aid to reflexive inquiry, improvisation and political adroitness. As such it relies heavily on practical judgment, a well-developed facility with group behavior and an ability to judge when to hold on-going conversation open and when it is necessary to reach temporary closure (Stacey, 2012: 121). Thus successful outcomes are not guaranteed, but it is my consistent experience that, with the caution outlined by Stacey, far more often than not it enhances shared understanding, coalesces shared will and provides indicators for meaningful action.

STRENGTHS

One of its main strengths is to discourage or interrupt linear thinking[2] and trigger emergent thinking, particularly and especially with what Dalmau & Tideman (2011) call "middle ground problems". By their nature they are both complicated and in many instances complex.

More formally they are those problems around which there exists only partial certainty as to their nature and hesitant levels of confidence as to the efficacy of any know solution set.

Practically speaking, when used as a facilitation tool for dialogue, a person can enter a room and discover, through the process of the conversation, new perspectives, thoughts, ideas and priorities they would not have previously imagined possible. Now many process devices can do this and some would say this is just simply "life", but what seems distinctive about the Process Enneagram is that invariably the new perspective is more creative, holistic and emotionally valent to the person involved.

Nevertheless attention does need to be given to the facilitation process to enable this to happen. There is no doubt it is a robust framework that can drive guidance of group dialogue, however when working with groups there is some further understanding and skill required to get beyond the predictable outcomes of linear reductionist thinking.

A WHOLE-OF-SYSTEM PERSPECTIVE

The range of applications of the Process Enneagram is limited only by the types of problems faced. For example, it is ideally suited to the aforementioned "middle ground" problem type. Such middle ground problems often present in the workplace as "people" issues involving emotions, cultures, diverse viewpoints, dynamic environments and marketplaces, or those that are just plain massive and complicated.

One key benefit of using the Process Enneagram to guide conversations around such problems is that it seems naturally to trigger a whole-of-system view of a situation. Whole-of-system viewpoints tend to be the exception rather than the norm in modern corporate life and yet are absolutely necessary for resolving complex prob-

---

2. Linear thinking—for the purpose of this paper linear thinking is referred to as such thinking that logically and predictably follows a set path of stable milestones toward a foregone conclusion. More formally it can be thought of as a process of thought following known cycles or step-by-step progression where a response to a step must be elicited before another step is taken.

lems of the middle ground.

For example, the Process Enneagram both evokes and allows for consideration of, say, implications not just for productivity, but also the workforce, culture, community, environment, industry, greater community, up to a national level if required and beyond - consideration both within and out. This tends to produce more comprehensive, informed and relevant decision-making and outcomes.

The power of the Process Enneagram as a process framework to bring a group together on social and emotional levels is not to be under estimated. Dalmau & Tideman assert the order of the Process Enneagram is instrumental in bringing about rational, social and emotional outcomes (Dalmau & Tideman, 2011). This is something I have found not only in theory but also in practice. Many diverse groups with contrasting ideas and diversity of thought can be brought together in a way that does not diminish the ideas of others. It is perhaps, in part, the inclusive, brainstorming nature of the earlier elements of the Process Enneagram framework where ideas and opinions need not be judged or discounted, but rather listed and listened to with some general concurrence. The frequent use of visuals helps participants detach emotionally from issues they would otherwise have had heated arguments about. The overall process framework tends to take some of the pressure off a group that would otherwise be naturally quite judgmental of each other - it allows them to just listen to each member of the group and what they have to say without critique. The critique coming later in the process (when discussing strategy and work) provides time for ideas to germinate, settle and be considered carefully, not hastily discounted.

APPLICATION

The Process Enneagram has many applications, including, but not limited to use as a tool for individual clarity and exploration of thought, as well as with groups as a facilitative or consultative tool. A client put it succinctly; *"From an individual standpoint, the PE helps me uncover things I hadn't thought of, generative avenues to explore. Not just by the systemic nature of how it looks at an issue or problem but also through the sequencing of the alternative viewpoints it encourages me to break free of the linear thinking that I was trapped in or unconsciously 'blocked' by when resolving issues."* For this reason, using the Process Enneagram on an individual basis can be a valuable resource to expand thinking, open doors and allow one to see different perspectives.

According to Richard Knowles, originator of the Process Enneagram as a tool for systematic and emergent problem-solving, the order of foci for the dialogue topics is important. This accords with my experience to a certain extent. Knowles (2012) recommends the order to be *Current State/ Identity*, *Relationships/Connections* and *Information/Will*, followed by *Intention*, *Principles/Ground-Rules*, *Tensions/Issues*, *Strategies*, *Work* and *Deep Learning*. Knowles[3] suggests this order because of a number of benefits that accrue ranging from promoting emergent thought to improving the social fabric of a group.

However it is my opinion and experience that this order can be successfully adjusted to context without sacrificing any of the benefits of using such a strict order. For example, with a group of people used to highly linear thinking, (often common among engineers or accountants when congregated in group settings), the order can just as easily evolve as *Current State-Intention-Tensions-Principles-Strategy* with *Relationships* and *Information* evolving as a matter of course throughout the dialogue, thereby remaining 'ecological' to the group.

It would seem that many in management roles are almost indoctrinated into, trained and rewarded for the linear thinking encapsulated by *'what's the problem, how do we fix it'*—treating this as an overarching mindset in their day-to day management and decision-making. For such people it can be extremely difficult to even talk about the *Current State* of a complex problem without highlighting the *Tensions and Issues* preventing them from achieving what they want or are tasked with achieving. Moreover it can be downright tortuous for some to talk about *Intention* without immediately jumping to the obstacles preventing that intention from occurring. This pattern appears more prevalent among middle and lower managers than executives in my experience, but nevertheless exists across the spectrum of management.

As in the myth of the ancient Greek god Herakles, life is meant to be a struggle. It is as if they have a goal in mind (although often times somewhat ill-defined) and then must also have a list of obstacles to overcome in reaching that goal. Then without thinking (or so it appears) their modus operandi is to make it their business and priority one of finding solutions to each of those problems, checking off their list.

This is fine for what Dalmau and Tideman (2011) refer to as bottom left hand corner thinking or what Stacey (1996) describes as ordinary management thinking. But this is not the type of situation for which the Process Enneagram is most useful. In fact, applying the Process Enneagram to these types of problems often complicates the issue and wastes time. Where the Process Enneagram provides real benefit is

---

3. Knowles in private correspondence has stated that his thinking on this has changed with time.

"middle ground" problems and for these it is invaluable.

Individuals caught (mostly) in linear thinking patterns often seem quite resistant to breaking this pattern initially. If the process used to facilitate dialogue around a complex problem is too different from the operating norm, attendant resistance can prevent engagement and can derail not only the dialogue at hand but also the wider acceptance of this type of inquiry in the organization as a whole. Like a dance where one leads and the other follows, it is my experience to follow more the natural inclination and order of the group at hand and once they have had a positive experience with a dialogue based on the Process Enneagram then be more inclined to follow the order as recommended.

Moreover, I have had consistent success in exploring *Relationships*, *Information* and *Identity* throughout the entire conversation by probing and questioning of all of the other 6 points and don't think they necessarily need to be addressed separately. Work in Australia, New Zealand, and the United States suggests asking people to talk pointedly about *Relationships* can be quite confronting, especially when most in the room will be in existing power-based hierarchical and dependent relationships with one another. Additionally for those new to a dialogue based around the Process Enneagram the requirement to trust, take risks, speak authentically and openly with colleagues and with (for many) a person in the role of facilitator they may not know: these matters can be quite challenging for some.

For these and other similar dynamics it can also be challenging to use the Process Enneagram as an aid to dialogue with a group or team of which you are the formal leader. The leader's past relationship with team members, their capacity to dis-associate emotionally and their competency in a broader set of leadership skills outside the realm of the Process Enneagram and group facilitation can often make for either success or failure.

Watching other facilitators using the Process Enneagram framework to guide dialogue, it becomes apparent that some seem at times confused about whether or not (and just how much) to explore past, current or future and ideal *Relationships*, *Information* and *Identity*[4]. Experience suggests it is more beneficial to explore the nature of the flow of *Relationships*, *Information* and *Identity* (IRI) as part of the current state conversation—indeed a whole Process Enneagram conversation can be had just exploring the *Current State*. Then, the remainder of the dialogue, particularly around *In-*

---

4. For many working with the Process Enneagram these three factors are simply collected together and called IRI—identity, relationships and information.

tention and *Principles,* can rightly also include a focus on future/ desired *Relationships,* and *Information* flow.

Typically past and present elements of IRI tend to arise when exploring *Tensions and Issues* and this seems quite appropriate. Future elements can be expressed through the *Intention* and *Principles,* thus suggesting it can be quite functional to simply and continually address IRI throughout the dialogue rather than as separate elements or "topics".

However, holding the dialogue to the broad sequence seems to reliably trigger shared meaning, new perspectives, clear foci for action, positive disposition among those involved and enhanced relationships.

The facilitation of any meeting begins the moment the decision is made to have the meeting. So it is with using the Process Enneagram framework and that which goes on in the manner of preparation is often as vital to success as what actually happens in the dialogue itself. Pre-work for success requires that one first ensure the problem under consideration actually is a complex issue, that is a middle ground problem. Dalmau & Tideman (2011) provide a checklist for identifying such issues. The problem needs to meet one or more of the following criteria:

- Be one where the situation is complicated;
- The change sought is complex;
- Outcomes are vague or unclear;
- There are unknown or unpredictable forces that can influence or interfere;
- People's feelings or reactions are likely to be triggered significantly;
- There is a need to equip, educate or train others to implement and sustain the change;
- There are politics involved or likely, and;
- Individuals or groups have the potential to feel disenfranchised as a result of the change.

It behooves the facilitator to be clear with the sponsor that the issue to be tackled really does meet one or more of these criteria. Otherwise frustration and impatience with redundant analysis inevitably attend such situations.

When introducing the Process Enneagram to others for the first time whether or not to show the diagram is worth considering—it can occasionally trigger some unusual responses. Without knowing the real history of the Enneagram, the particular form upon which the Process Enneagram is based, its evolution and application to various fields in different forms, the occasional individual may have a negative reaction upon first seeing the diagram. They can misinterpret it as having some kind of quasi-religious or spiritualist linkage—this is alien to how the framework was intended or described by Knowles, nor how my colleagues and I use it. As with many other group process challenges, the resolution is based on simply acknowledging resistance ahead of time and informing the group of the actual origins and evolution of the Process Enneagram.

But more even more basic is the question of whether to actually present the diagram to a group of people, versus use the framework in one's thinking as a private guiding map. Do people need to even see it in order to gain value from a dialogue on which it is based?

Personally I have found it most useful to withhold this diagram with people 'new' to the Process Enneagram, provide them the experience first, and uncover it only if the goal is to skill them in the actual use of it. It is a quite complex structure that may confuse individuals if there isn't sufficient time to explain it. Moreover it can take time and focus away from the discussion at hand. Therefore, I typically use it more as a mental map in my head that guides my process and informs the sequence of questioning that I use to take a systemic look at a given topic. The dialogue in the group is the main purpose, not becoming lost in a diagram or points of documentation.

It is not necessary for the facilitator to have an in-depth or detailed knowledge of the topic content when using the Process Enneagram framework. Experience, however, suggests the facilitator be somewhat conversant with the language of those who will be present in the conversation. By language I refer to the corporate 'lingua franca'. This ensures a standard that allows them to understand that which is being discussed, albeit in limited detail, and place group comments into their relative place on the Process Enneagram faster. This facilitator's basic understanding of the content and its context is important in determining whether the dialogue is consistent with the Process Enneagram framework or being sidelined off track.

Middle ground problems are by their nature complicated and not easily solved in one conversation. Experience suggests a good facilitation is based on planning for more time than you think you need and/or setting expectations so that participants understand it is a step in the right direction that will need further work—in reality this is the "spiral" nature of most human interaction and of the correct use of the Process

Enneagram. It is (and is intended to be) a process that is cyclical and thus open to be re-visited.

WHERE IT DOESN'T WORK SO WELL AND WHAT TO DO ABOUT IT

Experience suggests a common trap for new players (i.e., those new to facilitating dialogue using the Process Enneagram framework) is ritualistically following the advised order of topics and asking formulaic questions at each point in the prescribed sequence. Indeed the facilitation can, in such circumstances, actually promote linear thought and simplistic inappropriate solution sets. It almost encourages participants to become eager and insistent to "jump" straight from the issue or intention towards solution.

The cost of this occurring is the delay or outright prevention of the conversation exploring deeper underlying aspects of the problem, for surprise to arise from new information and (especially) for new perspectives and innovative solutions to emerge. Moreover, dissatisfaction with the framework itself and what may have offered the promise to all involved of better ways to converse about complex problems, then rises and the Process Enneagram is cast into that basket of "just another consultant's process", indistinguishable from all the other similar processes to which managers and executives have been exposed over their careers.

Then there is the question of just how, what and when to document. The use of visuals, e.g., whiteboards, flip charts, stabilizes and focuses attention in a group meeting. It is a vital component of the Process Enneagram's success for generating order out of chaos.

On the question of what to document there is a real tension on one particular point of the dialogue: in an effort to foster trust and openness in groups I occasionally find myself caught between scribbling down some of the solutions (which have been generated with somewhat simplistic linear thinking as the group traverses the early points of the dialogue sequence) and on the other hand facing the risk of losing permission with the group in the knowledge they will most probably develop far more complex insight later in the process.

The danger in not recording all suggestions visually is to risk losing some of the really important solutions or suggestions that may arise and yet by so doing, when it comes to discussing strategies and work, the points already mentioned can tend to send the group back into a simplistic linear mindset. It is around such matters as this that rests the real difference between the art and science of Process Enneagram facilitation. One resolution that seems to work well is to let the group know ahead of time and to request they hold off offering solutions till later. But even when done with a gentle, respectful and caring manner this too can shut down the group, constricting the flow of conversation that is so integral and valuable to the entire Process Enneagram process.

Another option is to record such suggestions in perhaps a different manner or at a different position from, say, the main record of the dialogue. More generally, success seems to come with acknowledging the solution and then moving the group along—a successful dialogue based on the Process Enneagram is often just about good group management and facilitation.

There are many ways to capture the output: when not in a training setting, I tend to document the dialogue on my laptop computer in real time onto a blank Process Enneagram template. This is a non-distracting way of recording the dialogue whilst simultaneously facilitating the conversation. It does however require a certain level of typing skill and solid familiarity with the Process Enneagram so that comments can be quickly put into various categories. At the end of the meeting these notes are distributed to the group in paragraph/ story format.

When training groups in how to use the Process Enneagram, I tend to record the core points in real time on a white board or flip charts in front of the group, and then photograph it for them to take away. It is important when doing this however that the group is having the discussion with itself, i.e., group members looking to each other and not to the facilitator or to the board—this way the dialogue flows and the full benefit is realized.

In terms of spatial arrangement, I typically prefer to seat a group in a circle without desks to encourage group engagement. Otherwise, the dialogue rarely goes as "deep" as it should and participants tend to stay more "in role".

Being clear as to one's role as a facilitator is important. Are you there to:

• Be in control;
• Get a result, or;
• Help others to explore their world.

All are important for they influence the choices a facilitator makes to guide the conversation through the structure and sequencing of the Process Enneagram. It is important the facilitator has clarity around their own personal outcomes, goals and ways of operating, separate from, but in alignment with the goals of the client for whom they are contracted to run the session.

I have observed many a new facilitator shut down a group through either the need to adhere strictly to the order of the Process Enneagram and/or by the manner in which they tried to control the group for some other reason e.g., limit off topic conversations, follow own agenda, display a bias towards certain individuals etc. These aspects of group handling, which some could consider quite basic aspects of leadership skill, can nevertheless easily derail a Process Enneagram dialogue: I do believe it should be led and guided by individuals with skill at group management and group dynamics.

Occasionally I come across individuals with a very low threshold for spending time in 'meetings' and conversation, as distinct from "action". The skill of the facilitator comes to bear in such matters, particularly around setting expectations: e.g., *"bear with me or trust me with this, this conversation may seemingly go in many different directions at times but I can assure there is an underlying structure."* It is important to follow this with reassurance that the group will get to an action plan and follow through with a strategy to create the intention.

WHO SHOULD USE IT

I have found the most effective use of the Process Enneagram occurs when the facilitator has had prior experience using it, is very familiar with the points and their order and so is readily able to track the dialogue from a process point of view.

Such a person is strategically better able to guide those unfamiliar with the process to a successful outcome. I would not recommend using the Process Enneagram if you are completely new to it (i.e., never experienced, witnessed or heard of it before) and the topic under consideration is both substantial and important to those involved.

AUTHENTICITY AND DURATION

My experience suggests the most successful method of achieving emergent thought comes not only through the order and roughly following the sequencing of the Process Enneagram points, but more importantly, through challenging the responses of those participating in the dialogue—hence the importance of one's skill as a facilitator in developing trust and a relationship with the group.

For example, consider the matter of *Intention*: many workplace groups will initially list off the 'usual suspects' when it comes to articulating what they want to achieve, i.e., world's best practice, world class leadership, leadership in safety, pre-eminence in the market. Awkward as it might be to state, these desires are often no different from dozens of other companies in the same field/industry and possibly geographic area.

Challenging a group to dig down deeper, getting to why "this" is important both individually and corporately, and uncovering values and beliefs is where, in my experience, both the buy-in and emergent thought really starts to takes place. It is too easy to rely on the Process Enneagram framework alone, to remain at a superficial level of conversation and therefore to not truly break free of linear thinking, to keep things "safe" for everybody.

Where the facilitator is very grounded as a person, is not hooked into the client's agenda or worldview, and takes a position of service, then deep authentic challenging can often trigger vastly different perspectives of the problem at hand, trigger motivating surprise and new forms of action.

I have generally found that half a day for most issues with a group of 6-20 people is not really enough time to get through many complex issues if one wants to go beyond the superficial level. Typically a diverse group will get some closure on the strategy element in half a day, providing a follow-up meeting is allocated to 'the work and deep learning'. This does of course however depend on the group and how aware they are of the topic under discussion, how much interaction they have already had around this topic and the like.

Half a day is often not long enough due to the time required for deeper questioning around what it is they really want and why: intention, principles and ground rules. If there is a preponderance of introverts in the group then it can take even longer for they may require more than the average time for reflection on new points that arise during the dialogue.

Overall, it is useful to set aside most of a full day, recognizing that subsequent Process Enneagram dialogues on aspects of the same problem will typically be much shorter.

All this said, it is also possible to get significant gains with the Process Enneagram framework in less than half an hour with a small group of people. This can occur provided that

1. Permission is evident and present with and between the facilitator and also with those who are in the group;

2. The Facilitator has sufficient understanding of the Process Enneagram and skill in group management to a level where they can quickly and effectively move a group through each point on the diagram, and;

3. The facilitator has effective questioning skills, able to get to what really matters for each group in a level of detail that is sufficient and helpful to the outcome.

Thus whilst a complete Process Enneagram discussion may take some time, it is still very useful in a limited time-frame.

DIAGNOSIS

Although not the main focus of this paper, it is worth noting that the Process Enneagram provides an excellent framework to understand, select and guide organizational interventions, that is, a whole-of-system diagnostic tool.

Too often, it seems, managers and teams address a problem by coming up simply with a *Strategy*, or worse, the *Work*. Now this works well as a method of approach when the problem is known, understood, black and white, and agreed by all, but it can be catastrophic for more complex or middle ground issues. Often the *Strategy* or action steps *Work* were initiated without discussion or clarity around the *Intention* (what they really want to achieve) and, more importantly, why they want to achieve this. By skipping so many of the first steps of the Process Enneagram framework, the underlying IRI foundations (i.e., *Relationships, Information* and *Identity*) are compromised, damaged or eroded to the point of significantly impacting the overall result, performance or profitability: recipe for failure. Without the insight that comes from self-organizing systems theory and without an understanding of the Process Enneagram framework, this looks to the casual observer as simply incompetent leadership of change or, even worse, top down command and control management.

This is where the Process Enneagram can add real value, i.e., as a model for diagnosis[5]. True, it is possible to use it as a way of understanding the *Current State* from a whole-of systems viewpoint, however when used backwards, it can be very effective at pinpointing the cause for a *Strategy* not working.

For example, when people are doing '*Work*' and not finding success, one can track back by asking them what their overall *Strategy* was, what the *Tensions and Issues* were that they had to overcome, and what the *Principles* were that they had agreed to do all work by. Bearing in mind that the foundations of *Relationships*, *Information* and *Identity* are visible throughout all of the other 6 points and do not need necessarily to be addressed separately unless the situation calls for it, or insufficient understanding has been obtained by exploring the other points.

The real power of this diagnostic approach lies when it is facilitated and done in a group setting. As the realization develops in the group or team of what has "gone wrong", a higher order of understanding *and* the real underlying problem tends to emerge, and if facilitated well guilt and blame never appear, motivation to find new and better perspectives develop. Moreover, it becomes a very useful but easy way of triggering systems thinking in a group of people without ever having to teach them about it.

REPRISE

In summary, the Process Enneagram is a very effective facilitative tool to use for a complex, middle ground problem, where a whole-of-system view is required and emergent thinking sought. It necessitates a certain degree of skill by the facilitator and basic group handling and questioning capabilities as well as familiarity with the process itself. Setting process expectations at the beginning of a dialogue will tend to make it run more smoothly, as will understanding the audience with which it is to be used. It can be used individually or with groups of various sizes for a range of applications and outcomes.

---

5. This particular and real benefit of the Process Enneagram was brought to my attention by Steve Zuieback (2012).

REFERENCES

Dalmau, T. and Tideman, J. (2011). "The middle ground: Embracing complexity in the real world," Emergence: Complexity & Organization, ISSN 1521-3250, 13(1-2): 71-95.

Knowles, R. (2002). *The Leadership Dance: Pathways to Extraordinary Leadership Effectiveness*, ISBN 9780972120401.

Stacey, R. (1996). *Strategic Management and Organizational Dynamics*, ISBN 9780273708117.

Stacey, R. (2012). *Tools and Techniques of Leadership and Management: Meeting the Challenge of Complexity*, ISBN 9780415531184.

Zuieback, S. (2012). *Leadership Practices for Challenging Times*, ISBN 9780983033615.

Chapter 5

HOLISTICALLY EDUCATING GRADUATE STUDENTS FOR THE CONCEPTUAL AGE USING THE PROCESS ENNEAGRAM©

H. Mark McGibbon

The author of this article addresses a graduate school exercise to educate students on how to think differently about problem solving and transformation. Students participate in a multi-phase classroom exercise that demonstrates their dependency upon a collective society. However, society is dynamic and complex and there is a requirement to introduce and apply a new 21st century tool or framework to assist in problem solving and transformation. The new tool is the Process Enneagram created by Dr. Richard N. Knowles. The Process Enneagram™ framework assists the author's graduate students in applying a holistic, agile, and dynamic to problem solving and transformation within the Information and upcoming Conceptual Age.

As a graduate school professor most of my students have a good understanding of technology. We discuss the changes in technology within their lifetimes by drawing a timeline, by decade, on the classroom board. Students remember and shout out their favorite technologies for me to categorize on a discreet 20th and 21st century timeline. This classroom exercise is fun, bounded, and simple. Students feel a sense of nostalgia and accomplishment for participating in the creation of a comprehensive technological timeline. This rewarding process demonstrates the students' ability to remember, understand, apply, analyze, evaluate and create in a collaborative environment.

The next step in this classroom exercise is for the students to research the process in which the technology originated starting from the idea stage to the final consumer. For timesavings sake, students form teams of eight team members to research a particular technology of their choosing. Their choice in teams is critical in creating a cohesive unit that has the desire to research a given technology with a common goal of exploration and discovery. During the students' research process, the students quickly realize the complexity involved with bringing one technological product to market.

To add to the complexity, students must answer several questions such as: What was the catalyst for technological change? Was the change sudden or incremental? Who were the major stakeholders? What were the major stakeholder's motivations for participating in this technological venture? Why did the end-users embrace the new technology? Was the technology part of a previous time-period and then reintroduced into the market under different market conditions? What changes in society or end-user demands caused the technology to be successful in a particular time-period compared to another time-period? These types of questions prepare the students to use critical thinking in a complex environment.

Once the students realize the complexity involved with introducing technology into the market place, students reflect upon how to improve the processes involved in knowledge acquisition, design, engineering, production, marketing, advertising, and distributing technology related products and services via another team exercise. Inevitably, the students commence writing linear formatted or sequential lists to decompose their task into small and manageable processes. A team representative from each team writes a sequential list comprised of multiple steps to create a means of conceiving new technological ideas for successfully introducing new technology into the market. When other professors and I introduce a problem into the student's sequential list of steps, often the students think of one or two ways to solve the immediate problem without thinking how the solution to their problem may affect other process areas.

At this point, some students begin to realize our 21st century means of problem solving in teams requires rethinking. Academia is to blame. From kindergarten to higher-education institutions, education professionals have placed academic subject areas in categories such as physics, chemistry, mathematics, discreet math, history, ancient history, English, U.S. literature, Russian literature, religious studies, mythology, architecture, theater, art history, U.S. history, computer engineering, electrical engineering, chemical engineering, business administration, business marketing, organizational leadership, etc. However, the human experience is a collection of these and other academic fields of study. If all humans studied only discreet math, where would humankind be today? Life is not strictly experienced in a sequential manner or one academic field of study. Humans depend on multiple experiences and knowledge basis to advance from good to great.

Life experiences and one's knowledge operate in a parallel manner. To provide a simple example to my students, I have my students imagine they are alone in a soundproof and light deprived room without any signs of any life. The room has no walls or ceiling; however, there is a chair for the student to sit on. The students now think they are free to experience nothingness. Yet, the students experience enlightenment yet again. The students begin to hear their breathing, their heartbeats, their digestive system, and other subtle bodily sounds. They begin to experience thoughts of how the chair was to become at this particular location and at this point of time. They also think of the floor came into existence. The smell of food introduced into this scenario by the professor makes the students think more of how dependent they are upon each other for survival ever since the *Stone Age*.

Humans have evolved through the Ages from the *Stone Age* to the *Bronze Age*, to the *Iron Age*, to the *Agrarian Age*, to the *Industrial Age*, and now to the *Information Age*. Humans evolved through these *Ages* because of our ability to collectively remember and understand past successes and failures; apply and analyze new theories and concepts that advance the human experience; and finally, evaluate the human experience to become increasingly more innovative and creative.

Humans will soon evolve from the *Information Age* to the *Conceptual Age* where innovation and creativity become the coveted attributes of the *Conceptual Age*. Yet creativity and innovation can be disruptive and complex within any human society. Inevitably, some humans will embrace the change from the *Information Age* to the *Conceptual Age*, and others will not. Those who become Luddites will not reap the *Conceptual Age* advantages such as those humans who did not embrace tools in the *Stone Age*.

Within the *Conceptual Age,* my students learn that both hemispheres of the human brain are critical for success. The left-brain hemisphere is responsible for analytical thinking for professions such as accounting and engineering. The right-brain hemisphere is responsible for holistic thinking or arranging smaller thoughts into a collective whole needed in leadership roles. The individual using both brain hemispheres will need to balance their individual brain hemispheres. However, this individual balance is not enough for the success of a society operating in the *Conceptual Age.* To achieve greatness in any society, the collective team, organization, society, nation, or world, must feel, think, and act together. Thus, the students become enlightened and understand that regardless of their individuality, humans rely on the collective team, organization, society, nation or world to sustain life.

Human life depends upon an amalgamation of human knowledge and experiences. Contrary to the characterization that Americans are individuals with free-reign to an open and free society, we have governing principles that apply to the collective society. Americans and other nationalities do not live or work in a vacuous bubble caring only for their own self-interests. Humans blend multiple facets of their complex lives together to create a more harmonious environment.

However, life, as students understand, is not always harmonious. Life is complex. Life is parallel. Humans depend upon other humans to survive and evolve. Humans have different roles in society. We work within a bubble, a *global bubble*. Without our collective knowledge, experiences, and insights, humans would cease to create and innovate. The bottom line is that we all assume roles in our *global bubble* and the *global bubble* is complex because of a myriad of human-related circumstances.

From this point in the lesson, the students yearn to solve the *global bubble's* many complex problems, but without an adaptable tool or framework. The students still cannot break the serial thinking process that society has taught them to embrace within their lifetimes. At this point, the students go on a 15-minute break to reflect upon the prerequisite lesson on the *Process Enneagram*™.

Returning from break, the students wait for me, as their professor, to solve the problems contained within the *global bubble*. Now it is time for another lesson. As we are all members of the *global bubble*, we must not try to rely solely on leaders (or professors in leadership roles) to solve humankind problems. Leaders provide guidance and mentorship from problem solving to setting the innovative environment in a complex and dynamic *global bubble*. This is the point in time to introduce the students to the *Process Enneagram*™. The *Process Enneagram*™ was the creation of Dr. Richard

N. Knowles and it is an adaptable and scalable framework encompassing multiple academic disciplines to assist small groups, teams, organizations, societies, a nation, and multiple nations solve problems and advance humankind via creativity and innovation.

Because many students have difficulty understanding a non-sequential framework such as the Process Enneagram™, I present what I refer to as the *Peanut Butter and Jelly (P&J) Sandwich Concept.* When a person squeezes a P&J sandwich, the jelly is the least viscous substance within the P&J sandwich, whereupon it oozes out the sides of the sandwich in unpredictable locations. The creator of the *Process Enneagram™* accounted for the P&J Sandwich Concept because life is unpredictable and when we push on one area of our lives; it may cause another area in our lives to react indifferently, violently, lovingly, etc. Students begin to understand the power of the *Process Enneagram™* because of its ability to *flex* in various directions, such as in life.

An introduction to the *Process Enneagram™* subject begins with what I coined the *Ouroboros Concept.* I explain the *Ouroboros Concept* as the perpetual renewal of life. We are born, experience life, die, and for some religious folks, we are reborn. The origins of Ouroboros come from the Greeks who display a snake in a circle that eats his own tail. The *Ouroboros Concept* as applied to the *Process Enneagram™* begins and ends with an identity. Usually the beginning identity needs changing. The end identity is usually a utopian or desired state. I stress the words *utopian state* to my students because perfection or utopianism is seldom, if ever, achievable; however, thoughts of utopianism provide humans with a quest to improve continuously.

Without delving into the details of the *Process Enneagram™* that are obtainable within the book, *The Leadership Dance: Pathways to Extraordinary Organizational Effectiveness* by Richard N, Knowles (2002), my students perform a final classroom exercise whereupon the students revisit their aforementioned technology research exercise. The students critically think how to transform a current day technological good or service into an improved or new technological good or service. Students must start using a familiar Enterprise Architecture (EA) approach in which there is a linear or sequential *existing state*, a *transition plan*, and a desired *end state.* Unlike the linear or sequential EA transition plan approach, the *Process Enneagram™* approach uses a rich, robust and flexing *P&J Sandwich Concept* process to fulfill the holistic *Ouroboros Concept of* letting go of the past to assume a new beginning in perpetual time. Once the students undergo this final exercise, many students experience a whole new perspective on leadership, change management, organizational behavior, societal norms, cultures, gender, policies, laws, procedures, social networking, data, information, knowledge, insights, governance, standards, work roles, business development, nation building, and the list goes on and on. The *Process Enneagram*™ assists students in understanding today's environment to shape tomorrow by taking into consideration the dynamic and complex world we all share within the *global bubble*. Therefore, the creator of the *Process Enneagram*™ revolutionized our 21st century problem solving thinking by enlightening us to think holistically about how to transform a team, organization, society, movement, nation or multiple nations within a complex *global bubble*.

REFERENCES

Knowles, R.N. (2002). *The Leadership Dance, Pathways to Extraordinary Organizational Effectiveness*, ISBN 0972120408.

Chapter 6

HOLISTICALLY EDUCATING GRADUATE STUDENTS USING THE PROCESS ENNEAGRAM©

Hua Wang

Grounded in the theories of complexity science, the Process Enneagram © was developed by Dr. Richard N. Knowles as a tool to help people in various social systems and organizations establish meaningful connections and work together effectively to accomplish common goals. In the last decade, the practical applications of the Process Enneagram in steel mills, coalmines, school boards, credit unions, law firms, and children's homes have demonstrated its significant value and transformative capacity for solving complex problems across the world. This paper contributes to the existing literature in two ways: (1) This is the first formal case study on the use of the Process Enneagram in the context of an institution of higher education; and (2) This particular application focused on the development of a value system and co-creation of an open and stimulating social environment rather than solving problems for people with already shared experiences.

INTRODUCTION

L earning takes place most effectively in a self-driven and socially stimulating environment. One of the biggest challenges in higher education is student engagement in the classroom. This paper reports the first formal case study on the use of the Process Enneagram © to help the university professor and students co-create a meaningful and shared learning space. I begin with a brief introduction to the Process Enneagram with its theoretical foundation in complexity science and development of the practical tool. I then provide the details on how the Process Enneagram was applied in a graduate seminar at a large public university in the northeast of the United States. I conclude this article with a summary of key lessons learned from this experience and a discussion on its further implications.

COMPLEXITY SCIENCE AND THE PROCESS ENNEAGRAM

C omplexity science emerged from the development in a number of natural science and later social science disciplines such as mathematics, biology, physics, chemistry, computer science, economics, sociology, and psychology. The convergence of the discoveries in these fields has led to a general approach to the study of *complex adaptive systems* (CAS)—collections of individual agents who have the freedom to act in unpredictable ways, and whose actions are interconnected such that one agent's action changes the context for other agents. This approach has been used to understand systems such as termite colonies, forest ecosystems, stock markets, immune systems, working groups, and networks of organizations (Goldstein, 1996).

Each term in the name of CAS points out a critical aspect of complexity science: *Complex* implies diversity among various elements of different characteristics. *Adaptive* suggests the capacity to change based on lessons learned from previous experience. And a *system* is made of interdependent and interactive single parts and it is more than just the sum of all (Lacayo, 2007). Therefore, the fundamental principles of complexity science emphasize that an organization is a living system not a machine; individual agents within the system can self-organize and make changes without central control; free flow of diverse information is essential for the system to evolve; order is emergent; the change process is nonlinear; and small changes can have big impacts (Capra, 1996; Wheatley, 1992). Grounded in the theories of complexity science, the Process Enneagram was developed as a practical tool to (1) help people in various social systems and organizations solve complex problems, (2) establish meaningful connections, and (3) work together effectively to accomplish common goals. The Process Enneagram is known to be the only tool to fulfill all the three-abovementioned purposes simultaneously.

Although the enneagram was first conceived by a Russian spiritual teacher Gurdjieff, introduced to the West in the early 20th century, and later elaborated in the writings of Ouspensky (1949) and Blake (1996), it had limited real world applications until the development and publication of the Process Enneagram (Knowles, 2002). Visually, the Process Enneagram is a diagram that consists of nine points evenly placed on a circle and connected by different types of lines (see Figure 1). These nine points are marked with numerical and corresponding textual labels (0—Identity, 1—Intention, 2—Issues and Ambiguities, 3—Relationship, 4—Principles and Standards, 5—Work, 6—Information, 7—Learning and Potential, 8—Context and Structure, and 9—Identity). The different types of lines indicate different processes. There are three primary processes: The Circular Process (marked by the circular solid grey line through points 0, 1, 2, 3, 4, 5, 6, 7, 8, 9), the Zigzag Process (marked by the solid red arrows from points 1 to 4, 2, 8, 5, 7, and back to 1), and the Triangular Process (marked by the dotted black line through points 0, 3, 6, 9). The Circular Process represents the elements visible to the agents in a CAS. It is the starting process and helps establish baseline data. The Zigzag Process represents the unfolding of a world that is invisible to most people. It is the process of transformation. The elements on the right side of the Process Enneagram can help agents in a CAS co-create a shared value system and an open and stimulating environment. The elements on the left side of the Process Enneagram can guide the agents to realize the values in their actions and accomplish their common goals. The Triangular Process represents the process of self-organization. It is the core process of the Process Enneagram because it is the source of meaning and integrity. The Process Enneagram helps people set boundaries, initiate self-organization and meaningful communication, uncover the true challenges that underpin the complex

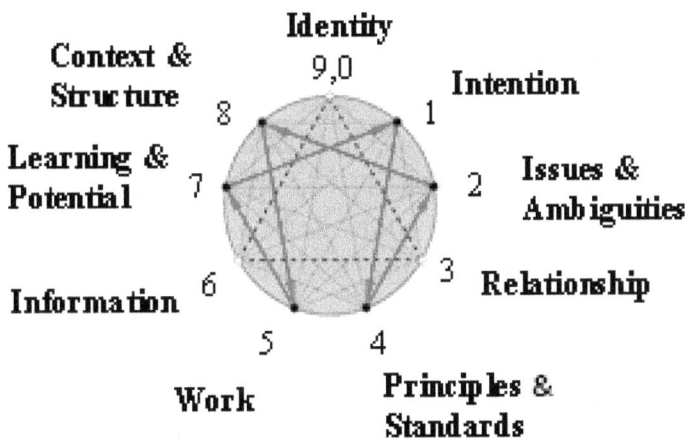

Figure 1 *The Process Enneagram*

problem they are facing so that they can establish open and trusting relationships, co-ordinate and collaborate to create effective and sustainable solutions (Knowles, 2002).

From 2002 to 2012, the practical applications of the Process Enneagram in steel mills, coalmines, school boards, credit unions, law firms, and children's homes have demonstrated its significant value and transformative capacity for solving complex problems across the world. This paper contributes to the existing literature in two ways: (1) This is the first, published formal case study on the use of the Process Enneagram in the context of an institution of higher education. (2) Reports on the use of the Process Enneagram to this date have been predominantly focusing on solving problems for people with already shared experiences. The case presented here rather focuses on the development of a value system and co-creating an open and stimulating social environment for a group of people without preexisting relationships or shared experiences.

APPLYING THE PROCESS ENNEAGRAM IN A UNIVERSITY CLASSROOM

In Spring 2011, I applied the Process Enneagram in a graduate seminar on Entertainment-Education that I taught in the Department of Communication, University at Buffalo, The State University of New York. A total of nine graduate students enrolled in the seminar, including five Ph.D. students and four M.A. students, eight from the Department of Communication and one from the Department of Geography, five were male and four were female, three were American students and six were international students. Most of them had not met before this class.

Introducing The Process Enneagram

The Process Enneagram was used as an icebreaker activity on the first day of class, January 19. After showing a TED talk by Sugata Mitra on Children-Driven Education, the class had a short discussion about how education is a self-organizing system and learning is an emergent phenomenon. Then the attention was brought to the diagram drawing on the foam board taped on the back of the classroom wall. The students were told that this diagram is called the Process Enneagram and it was developed by Dr. Richard N. Knowles based on his work experience in various organizational settings and his understanding of theories of complexity science, self-organization and leadership. The students were also informed that the Process Enneagram had been applied in different contexts to help people understand how activities take place and create an open, effective, and sustainable social environment to work together and achieve their common goals. Then the students were invited to participate in a class activity that used the Process Enneagram to explore a research question: "How do we make this class the best it can be?"

Running The Process Enneagram

Given that the students and I did not know much about each other, a combination of the Zigzag Process and the Triangular Process were adopted in this context and the class went through the elements together (following the points 0—1—4—2—8—5—7—3—6). For each element, I first explained what it meant, provided specific examples, and then invited the class to share their thoughts and questions. As our discussion went on, I made notes on the board and color-coded them to correspond to the specific elements (see Figure 2).

Identity: Who Are We?

I opened up and introduced myself first: "I am the instructor and I would also see myself as a facilitator and a participant in the process of learning and communication. I am a new faculty member in the department. I am passionate about the subject of this seminar entertainment-education and am very happy to share with you the exciting practice and research in this field." I also shared a short personal story about how I became interested in entertainment-education. Subsequently, all the students were invited to share a few things about themselves and how they see themselves as related to the theme of the class. Many students pointed out their interests and work experiences related to entertainment-education such as working with a NGO to promote organ donation, participating in anti-smoking campaigns, broadcasting radio programs, and organizing events at fan clubs.

Intention: What Would We Like To Have Accomplished In The Next 16 Weeks?

I distributed course syllabus and used this element to propose and discuss specific goals with the students. Some goals raised by the students were general (e.g., self-discovery, peer-to-peer learning) whereas others were more specific (e.g., Why American shows are so popular? How to use the entertainment-education strategy to train other educators? How can we work with media industry practitioners?). But all of them turned out to be very useful.

Principles & Standards: What Should Be Our Ground-Rules To Achieve The Goals?

We talked about the use of technology such as cell phone and laptop during class meetings. I told the students I would expect everyone to take the responsibility of doing the best we could. Sharing was important. And there were no stupid questions. I then asked the students if they had any suggestions. The issue of self-plagiarism came up. I had a chance to clarify the school policy. We also agreed on having a 15-min break and the boundary of appropriate media content to show for class discussions.

Issues & Ambiguities: What Challenges Might We Encounter In The Coming Weeks?

I brought up the issues of time management and possible challenges with severe winter weather and asked the students what problems they thought we might have. They proposed to organize a Skype conference call in case of severe weather. I also got to clarify the different purposes and requirements for some of the assignments.

Structure & Context: What Are The Physical And Social Environments We Are In?

I gave examples of the institutional context and the class schedule and opened it up to the students for further thoughts.

Work: What Are We Going To Do In This Class?

I went over the course materials, readings, assignments, research paper, and course website. I then asked the students if there were anything they wanted to add, remove, change, or clarify. I was glad to be able to incorporate the specific class information into this process and made small changes to respond to the students' requests.

Learning: What Are We Going To Learn? How Are We Going To Learn It Together?

Since it was only the first class, we didn't spend much time on this element.

Relationship: What Would We Want Our Relationships To Be Like In This Class?

I proposed the characteristics such as openness, mutual support, and trust.

Information: How Would We Want To Create And Exchange Information In This Class?

I suggested the use of textbooks, course readings, and course website to begin with.

Overall, the first class guided by the Process Enneagram went smoothly. It did require some preparation such as checking out the classroom space for posting the diagram and arranging the chairs and finding appropriate materials to post the diagram. Although the Process Enneagram is usually presented on flip charts in workshops, I was able to obtain foam boards and use double-sided tapes to post the diagram in the back of the classroom. The students and I spent more time discussing the elements on the right side of the Process Enneagram, which was consistent with our

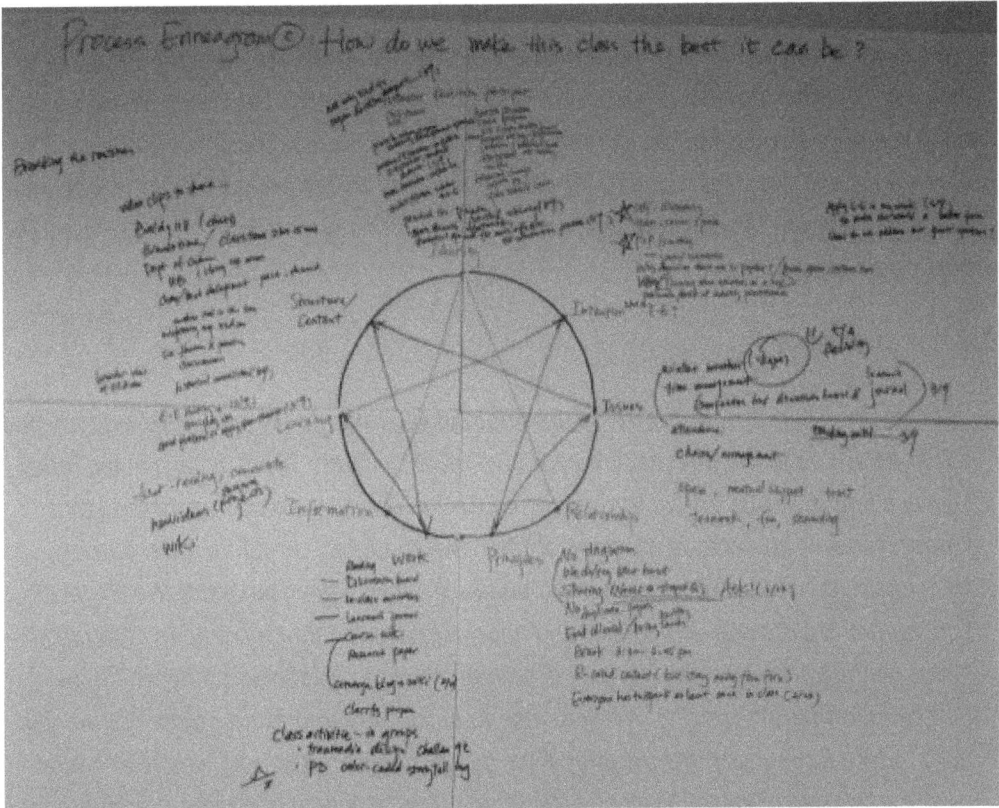

Figure 2 *Results from the Use of Process Enneagram*

purpose to establish a shared value system. My conscious use of the words such as *we, our,* and *us* as opposed to *I, my,* and *me* also helped cultivate a collective learning space. One student even pointed it out in a casual conversation afterwards that "you would say it is OUR class." Noting down the key words from students' comments and suggestions on the diagram board was an effective way to visually acknowledge their participation and appeared to be empowering, as this time around it was the instructor, not the student, who was taking notes.

Initial Feedback On The Use Of The Process Enneagram

As part of the class requirement, the students were asked to submit learner's journals each week. Although the specific topics for their self-reflection were not assigned, some students found the use of the Process Enneagram inspiring, personally relevant, and motivating. For example, one student commented,

The Process Enneagram © is a technique that is also relevant to self-organizing systems so I found it an especially interesting way to begin the course. I appreciated the opportunity to learn about my classmates and the instructor. The exercise made me

realize how much I can learn from everyone in the class this semester. I also appreciated the chance to discuss the times in which the learner's journals are due and the instructor's understanding of my educational and professional obligations. When I first reviewed the syllabus I had hesitations about taking this course (based on the quantity of readings and exercises required for a seminar), but I do believe now that I will be able to derive sufficient benefit from learning more about this subject and writing a research paper on a topic of interest.

Another student said,

"If he [the teacher] is indeed wise he does not bid you enter the house of his wisdom, but rather leads you to the threshold of your own mind" (Gibran, 1923). As explained today, I believe that this class on entertainment-education involves a teaching method that indeed leads the students to the peak of their own thoughts, stimulate them to think for themselves, and encourages them to build self-knowledge. The use of the Process Enneagram in class was very interesting and well prepared that inspired my thoughts about methodology building for the creation of effective social environments. I think that by following this process, we were able to tackle very important points that we will have to face throughout the semester. First, by letting us present who we are and why we are interested in entertainment-education, we were able to understand each other and have an idea of what to expect from each other. Also, we were able to learn others' ambitions, interests, and expectations in class. Second, we established some principles and rules such as the avoidance of plagiarism, the time for break, and many others. I am personally glad that I learned that using work from class for publication is not considered plagiarism. I have a great amount of well-organized work done in class since last semester that I would like to include in my future publications. Also, considering the structure of the class and the issues we might face during the semester such as weather conditions was very helpful in predicting possible obstacles with which we might have to deal with. Finally, it is inevitable to say that this class will be at the same time challenging, entertaining, and educational. It will create a self-organizing system that will progress throughout the semester on its own. In fact, the course will surely give students coming from diverse backgrounds and disciplines the opportunity to contribute to the entertainment-education strategy as they are using the entertainment-education strategy. The preparation for class through the discussion board, the in-class activities, the learner's journals, and the course wiki are expected to create the best learning environment for emergence to occur.

Finally, a student wrote,

The Process Enneagram seemed effective at navigating the learning and problem solving experience. The learning experience was largely an introduction to entertainment-education and the problem, so to speak, was to how we were going to engage in the learning experience in the class. ...The enneagram was part of our learning process in this first class. As we moved around to the different components we were learning, getting to know each other and setting up our classroom space. We were setting some structure and form to our process and social engagement in the class, led by the instructor. Overall I took our first class to be a very successful learning and problem solving experience.

REVISITING THE PROCESS ENNEAGRAM

I posted the Process Enneagram in the back of the classroom each week although we didn't necessarily discuss it every single time (see Figure 3). Instead, the graduate students and I revisited the Process Enneagram in the middle and towards the end of the semester to discuss changes in the past few weeks and new challenges and solutions. For example, on February 23, the students voluntarily suggested adding to the "Principles & Standards" that whenever someone has a question, they should just ask (similar to the ground rule of "no questions are stupid"; and everyone should speak at least once in class. These suggestions emerged through the classes in response to some of the cultural differences amongst the students as some of them came from Asia and were more hesitant to pose questions or make comments during class discussions. This suggests that the learning environment was safe and comfortable enough for their peers to encourage them voice their opinions. On March 9, some of the students saw a closer connection to the class topic and added to the "Identity" element that they have changed from passive receivers to active producers, senders, and receivers in the process of entertainment-education, and they could see potential research trajectories in this area. They also expressed for "Intention" that they wanted to apply the entertainment-education approach in their work to make the world a better place and had started to think about how to address their personal questions to the guest speakers I arranged towards the end of the semester. For "Issues & Ambiguities" we clarified the confusions between two class components and worked out a way to keep track of individual contributions on the course wiki. Finally, on May 4, which was the last class meeting for the guest speakers' visit and final presentations, the students said they really appreciated the fact that in our first class we raised the potential challenge with severe winter weather and suggested coordinating a Skype conference call. They were grateful for the flexibility and it actually proved to be an effective solution when a graduate

Figure 3 *Process Enneagram Diagram on the Back Wall of the Classroom*

student had to Skype in from his home in Rochester due to a snowstorm and road clo-sure. For the "Work" element, students suggested to add class group activities to the di-agram. They enjoyed the exercises to work in groups on a transmedia design challenge and using color-coded storytelling to understand a new concept Positive Deviance.

COURSE EVALUATION ON THE USE OF
THE PROCESS ENNEAGRAM

At the end of the semester, students were invited to participate in an anonymous course evaluation online. The results yielded much higher ratings of the course and me as an instructor when compared to my colleagues in the department and across the university. When they were asked to comment on the element of the course they found particularly effective, one student wrote, "The participatory ap-proach of the class is what makes it most effective!" and another said, "Cutting-edge methods made for an inspired pedagogy and a welcoming space for learning." When they were asked to comment on how effective the instructor was in teaching this course, one of the most flattering statements I received was:

"Opened new ways of thinking for me...was supportive of diverse research styles and methods...allowed the class to belong to all of us in a way that was engaging and respectful. Running a class is not easy, but she made the sessions something to look forward to each week and the readings were a perfect backdrop to understanding the subject AND to applying it for future research."

LESSONS LEARNED AND FURTHER IMPLICATIONS

Overall, it was a successful experience applying the Process Enneagram in the graduate seminar with my students. By focusing on the elements highlighted in the diagram and following the directions in the Zigzag Process and the Triangular Process, we co-created a shared, open, and stimulating learning space. Together, we transformed the classroom from a typically hierarchical place to a collaborative environment. Reflecting on the process, there were a few observations that I found important in this case study:

- The genuine invitation: When students see the instructor not only as an subject matter expert dictating their learning process in the class but rather candidly inviting them to take initiative and share along the way, the door is open and the shared space is in place.

- The surprising discovery: When the instructor and the students all embrace the diversity residing in themselves and value the knowledge, skills, and experiences they each bring to the learning process, they realize they can learn from each other and there are always new things to discover in each class. Learning becomes more fun.

- The empowering visualization: When students see their ideas on the wall, they tend to be more engaged because they feel a sense of ownership.

- The mindful arrangement: Gareth Morgan has a quote: "Farmers don't grow crops. They create the conditions for crops to grow." When the instructor is aware the importance of self-organization and makes an effort to create conditions to facilitate the learning process, new order will emerge and the students will flourish.

There are different approaches to learning (see Figure 4 from Plexus Institute). Very often, the experience becomes a one-way flow of distributed information from the instructor. More enthusiastic instructors may manage an interactive two-way communication with individual students, but it has been challenging to develop a learning space for a complex community with all-way collaboration among participants, including both the instructor and all of the students. This is where novelty and emergent understanding will coevolve. This paper reports the first formal case study on the use of the Process Enneagram in higher education. This experience as described and analyzed above suggests that the Process Enneagram is a powerful tool for fundamental transformations and has great potential for applications in educational settings.

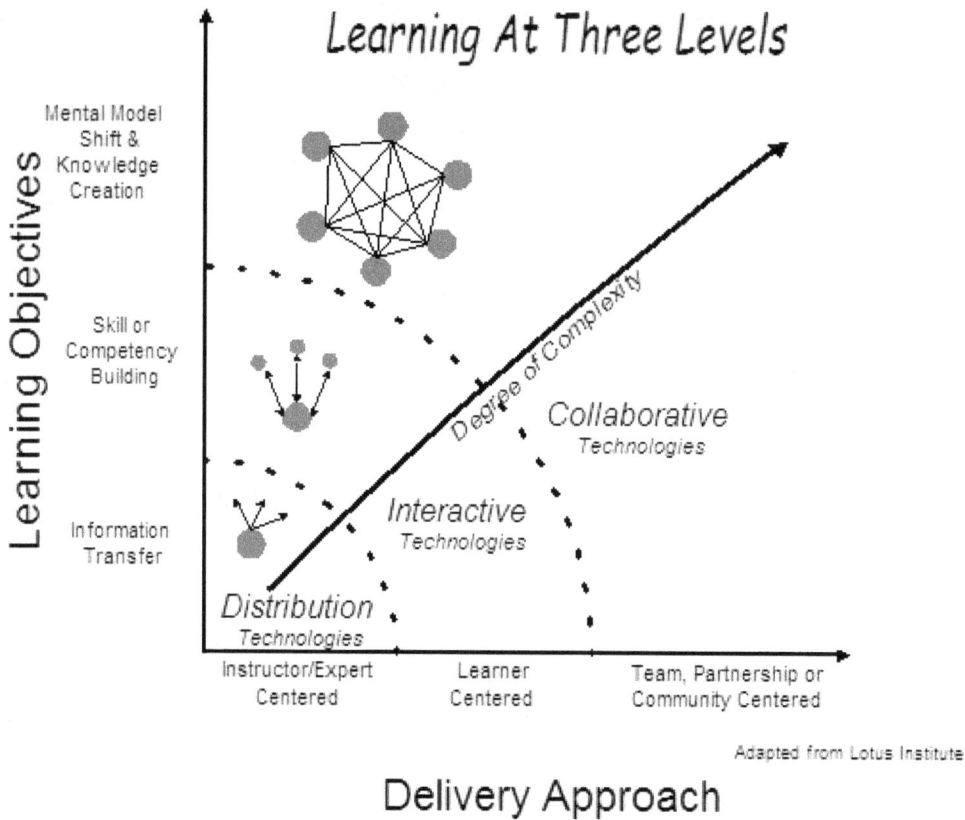

Figure 4 *Comparison of Learning Styles from Plexus Institute Introductory Course Material on Complexity Science*

ACKNOWLEDGEMENT

The author thanks Dr. Richard N. Knowles for his generosity in sharing his knowledge and experiences of the Process Enneagram and for his support of the development of this paper.

REFERENCES

Blake, A. G. E. (1996). *The Intelligent Enneagram*. Boston, MA: Shambhala.

Capra, F. (1996). *The Web of Life: A New Scientific Understanding of Living Systems*. New York: Doubleday.

Goldstein, J. (1996). Causality and emergence in chaos and complexity theories. In W. Sulis and A. Combs (Eds.), *Nonlinear Dynamics in Human Behavior* (Studies of Nonlinear Phenomena in Life Sciences, Volume 5), pp. 161-190, Singapore: World Scientific Publishing.

Knowles, R. N. (2002). *The Leadership Dance: Pathways to Extraordinary Organizational Effectiveness.* Niagara Falls, NY: The Center for Self-Organizing Leadership.

Lacayo, V. (2007). What complexity science teaches us about social change? MAZI Articles, Communication for Social Change Consortium. Available at http://www.communicationforsocialchange.org/mazi-articles.php?id=333

Ouspensky, P. D. (1949). *In Search of the Miraculous: Fragments of an Unknown Teaching.* New York, NY: Harcourt Brace Jovanovich.

Plexus Institute. *Simplicity on the Other Side of Complexity: An Introduction to Complexity Science and Management.* Retrieved on July 19, 2012 from http://www.plexusinstitute.org

Wheatley, M. (1992). *Leadership and the New Science.* San Francisco, CA: Berrett-Koehler.

Chapter 7

TOOLS OF COMPLEXITY: THE PROCESS ENNEAGRAM©

Barry W. Stevenson & Paul Rowland

This paper describes a three year journey using the Process Enneagram© as a tool for team development (Knowles, 2002). The director and eight managers of a mid-level management team associated with a large Canadian Telecommunications Company contracted an organizational development consulting firm to help bring the team closer together. The goal stated by the director was, "to enhance the capacity of the people that report to him". Over several years the mid-level management team grew stronger and became aware of the inter-relationship of each of the nine Process Enneagram components addressing various issues that emerged. Success in year one prompted each manager to sweep-in their respective regional teams. This examination and assessment of the application of the Process Enneagram as a tool for team development provides key insights on how best to engage with the client and set realistic goals for success.

IN THE BEGINNING

As an Organizational Development (OD) consultant it is often difficult to look back several years and examine what occurred when you were working with a complex organizational system like a team. So much of what you and others might have perceived is constricted by one's worldview and perceptions of reality. One of the ways consultants address this challenge is to talk with those who were actually involved at the time and seek clarity and consensus of actual events. In this paper, both the consultant and the client have come together to share their perceptions and reflect on what they understood to be factual; not just a matter for interpretation. In the beginning there was a request by the client to examine the way his team was operating with the stated goal "to enhance the capacity of the people that report to him". Thus began a three year journey which involved the use of the Process Enneagram as a tool for team development. This paper presents a shared reflection of events which over the course of three years lead toward significant improvements in team awareness and functioning and raised important questions regarding the application of the Process Enneagram as a tool for team development.

In order for the reader to fully understand the manner by which the Process Enneagram was employed, reference to the work of Richard N. Knowles is recommended.[1] The following is a brief summary of its application to the work presented in this paper.

The concept of self-organizing leadership has been applied, using the Process Enneagram developed by Richard N. Knowles. This process, in a very systemic and unique manner, integrates all of the essential components found today in team development and team building literature into a holistic approach to developing and maintaining effective teams. What the Process Enneagram approach offers is a way for teams to 'see' themselves as they emerge and evolve in the daily practice of being a team. Using this approach, a 'mirror' is created and continually adjusted to reflect the degree to which teams are actually functioning, based on the nine component elements that comprise the Process Enneagram. These are as follows: identity, intention, issues, relationships, principles/standards, the work, information, learning and structure/context. Representation of these nine component elements in the Process Enneagram, as developed and presented by Knowles, is as follows.

---

1. Richard Knowles is the founding director of the Center for Self-organizing Leadership based in Niagara Falls, New York. He has copyright to a nine-point enneagram process used in working with self-organizing systems and may be contacted through his company, Richard N. Knowles & Associates, Inc. Tel: (716) 622-6467 or through his mailing address at 6083, Bahia del Mar Circle, Unit 564, Saint Petersburg, FL 33715, USA. His e-mail is rnknowles@aol.com and his website is www.rnknowlesassociates.com.

The Living Systems Patterns and Processes

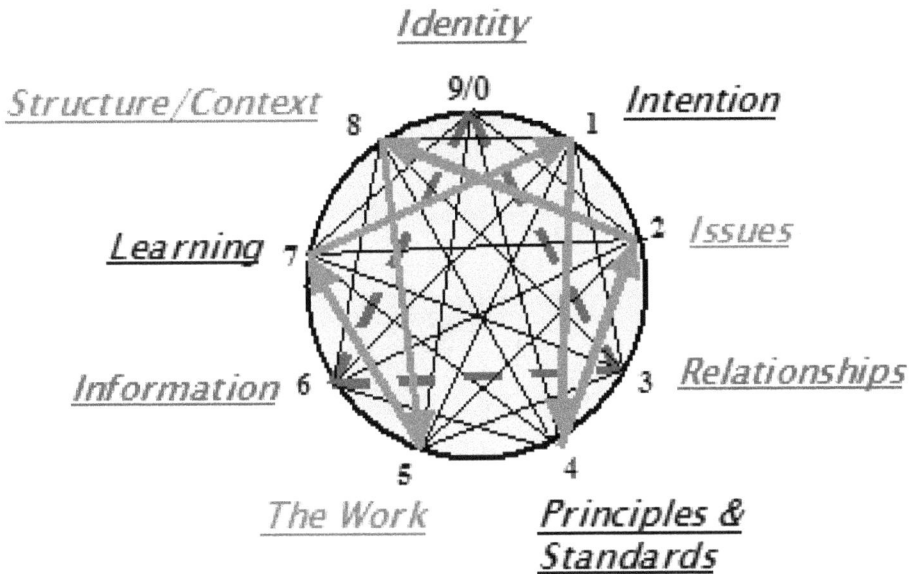

Figure 1 *The Process Enneagram*

Upon simple reflection, one can quickly see that all of the key points required to be addressed in team development work are found in these nine component elements. What is quickly apparent, however, and found lacking in more traditional team development approaches, is the inter-relationship and connectivity that is present in the way the Process Enneagram is described and used. Every component element is connected to every other component element. When one component changes, all of the others change. This 'dynamic' mimics the processes and patterns seen in teams as they function (or not).

We can quickly discover this dynamic by taking any simple relationship among the nine component elements and describing what happens when one thing changes. For example, when environmental changes force team members to readjust the way they have been doing some task together, new relationships and ways of sharing information emerge. Learning from these new relationships is often followed by revision in the work undertaken, the structure and context of the work itself and the way issues are addressed. The team's sense of identity is impacted, as new ways of 'being' together require a re-grounding in the norms (i.e., principles/standards) supporting the various relationships emerging and being established within the team. As suggested, no one component element can change without affecting all of the others. In this way, the Process Enneagram represents how the team, functioning 'as a living system', actually works. By doing so, it both establishes a pattern recognition process

Map 1

How do we currently function as a team?

Structure of Organization/Division.
Command & Control
Hierarchical/ Bureaucratic
Very Departmentalized/ Silos

Structure of Team
Consensus approach/Democratic
Shared leadership
Leadership focus on What, not how

Context Overall
Telecommunications
Highly regulatory/Highly competitive
Increased consumer interaction
Innovative/High tech.

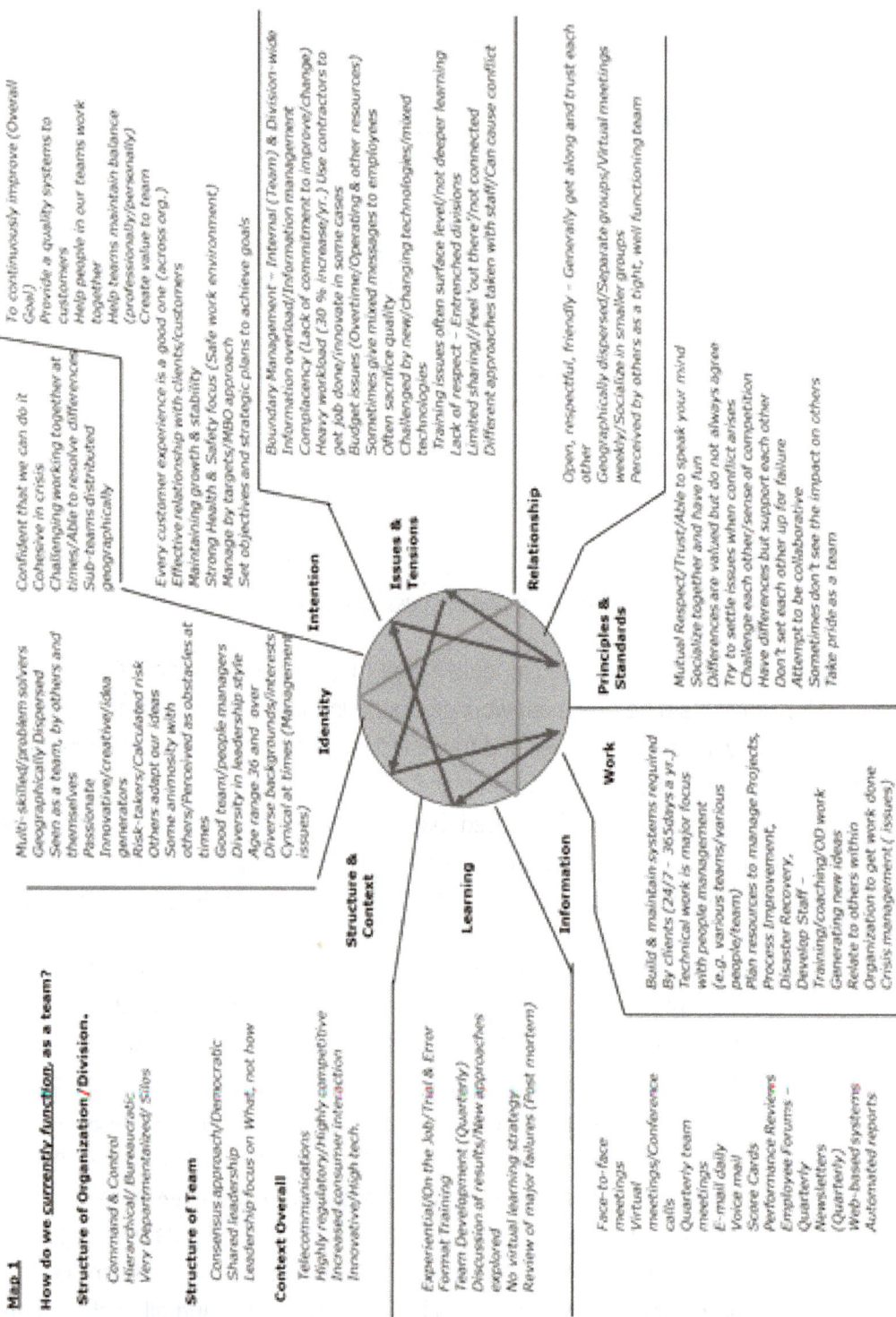

Learning
Experiential/On the Job/Trial & Error
Formal Training
Team Development (Quarterly)
Discussion of results/New approaches explored
No virtual learning strategy
Review of major failures (Post mortem)

Information
Face-to-face meetings
Virtual meetings/Conference calls
Quarterly team meetings
E-mail daily
Voice mail
Score Cards
Performance Reviews
Employee Forums – Quarterly
Newsletters (Quarterly)
Web-based systems
Automated reports

Identity
Multi-skilled/problem solvers
Geographically Dispersed
Seen as a team, by others and themselves
Passionate
Innovative/creative/idea generators
Risk-takers/Calculated risk
Others adopt our ideas
Some animosity with others/Perceived as obstacles at times
Good team/people managers
Diversity in leadership style
Age range 36 and over
Diverse backgrounds/interests
Cynical at times (Management issues)

Intention
Confident that we can do it
Cohesive in crisis
Challenging working together at times/Able to resolve differences
Sub-teams distributed geographically
Every customer experience is a good one (across org.)
Effective relationship with clients/customers
Maintaining growth & stability
Strong Health & Safety focus (Safe work environment)
Manage by targets/MBO approach
Set objectives and strategic plans to achieve goals
To continuously improve (Overall Goal)
Provide a quality systems to customers
Help people in our teams work together
Help teams maintain balance (professionally/personally)
Create value to team

Issues & Tensions
Boundary Management – Internal (Team) & Division-wide
Information overload/Information management
Complacency (Lack of commitment to improve/change)
Heavy workload (30 % increase/yr.) Use contractors to get job done/innovate in some cases
Budget issues (Overtime/Operating & other resources)
Sometimes give mixed messages to employees
Often sacrifice quality
Challenged by new/changing technologies/mixed technologies
Training issues often surface level/not deeper learning
Lack of respect – Entrenched divisions
Limited sharing//Feel 'out there'/not connected
Different approaches taken with staff/Can cause conflict

Work
Build & maintain systems required
By clients (24/7 – 365days a yr.)
Technical work is major focus
with people management
(e.g. various teams/various people/team)
Plan resources to manage Projects,
Process Improvement,
Disaster Recovery,
Develop Staff –
Training/coaching/OID work
Generating new ideas
Relate to others within
Organization to get work done
Crisis management (issues)

Relationship
Open, respectful, friendly – Generally get along and trust each other
Geographically dispersed/Separate groups/Virtual meetings weekly/Socialize in smaller groups
Perceived by others as a tight, well functioning team

Principles & Standards
Mutual Respect/Trust/Able to speak your mind
Socialize together and have fun
Differences are valued but do not always agree
Try to settle issues when conflict arises
Challenge each other/sense of competition
Have differences but support each other
Don't set each other up for failure
Attempt to be collaborative
Sometimes don't see the impact on others
Take pride as a team

Center nodes: Structure & Context · Learning · Information · Identity · Work · Intention · Issues & Tensions · Relationship · Principles & Standards

Figure 2 *Current Team Function*

for monitoring team functionality and provides an accountability structure for holding team members accountable to each other in their work.

Returning to the focus of this paper, what follows is a description of a three year journey using the Process Enneagram as a tool for team development and the outcomes associated with this work. The director and eight managers of a mid-level management team associated with a large Canadian Telecommunications Company contracted with an organizational development consulting firm, namely the co-author[2], to help bring the team closer together. As the team development work commenced it was clear that one of the main objectives set by the client was to build more effective relationships among team members and between the team and its various stakeholders. It was also essential to fully understand the nature of the team as a complex adaptive system as reflected by the nine Process Enneagram components outlined earlier.

The way chosen to initiate this was through story-telling and dialogue with team members, followed by asking the question, "How do you currently function as a team". The 'initial Process Enneagram inserted below in edited version[3] shows how the team generally saw itself functioning as a team

The 'picture' presented in this initial map shows the team functioning not unlike any team operating within a larger organizational structure. The team saw itself as generally together on things that needed to be done given the overall intention (i.e., goals/objective) for the team compared with how it saw the work it had to do. There was a sense of camaraderie among team members, not unlike an all-male sports team where the 'guys' get the job done and take pride in doing so.

On the surface it looked great. But what underlay this euphoric and at times macho atmosphere displayed itself in at least two important ways. First, there were some cracks in relationships among some members and between the team and other parts of the organization. When pressed to be specific team members tended to generalize. Second, there was a 'we're just fine' attitude that suggested more going on under the surface of the discussion.

After letting this initial map present itself to the team members as they reflected on the work done over a lunch break, the afternoon began with a feedback session on Map 1 which challenged team members to be as open and as trusting as possible

---

2. B.W. Stevenson & Associates, Ltd. #1603—838 West Hastings Street, Vancouver, B.C., Canada, V6C 0A6.

3. Each of the three *Process Enneagram* 'maps' presented in this paper have been edited to remove any direct reference to the actual client for confidentiality purposes.

in looking at a second question, namely "How do we want to function as a team?" As the Map 2 shows (Figure 3 bold type), there was considerable energy devoted to principles and standards.

As the team continued to work towards improving their team dynamics they began to see the value of setting objectives around what were termed the 'big rocks'. Team members began to talk differently, more openly and with a deeper sense of intention and commitment to improve themselves both in addressing areas for action and in living the principles and standards espoused. This initial team development work set in motion a palpable dynamic and collective energy for future action and improvement. Map 2 (Figure 3) was copied and sent to each team member to display on their respective office walls as the team director modeled. Each team member, to various degrees, took it upon themselves to share this work with their respective teams in the field. In receiving feedback several weeks and months after the mid-level management team development session it was evident that the team development work based on the Process Enneagram had allowed this team to better understand their current situation and helped them clarify and set objectives for improvement. One might say they were on a roll at this point.

Over the following year the OD consultant kept in touch with the team director (i.e., client) to monitor progress. With the intent of furthering this work throughout the various regional teams represented by each manager on the mid-level management team, the OD consultant was asked to facilitate workshops using the Process Enneagram process for all but a few of these regional teams. Those that the consultant did not facilitate were addressed by the appropriate team managers, based on their respective experiences with the initial team development process.

In facilitating six of the nine regional teams, the OD consultant observed that the application of the Process Enneagram had a variety of effects and outcomes. What these facilitated team sessions held in common was a commitment to work through the process, as the various mid-level management team members had done. The various regional team members displayed significant interest and committed effort concerning components such as shared principles and standards (i.e., shared values), improved relationships, better information sharing and clarity regarding the work required to meet the overall intention (i.e., purpose) of the mid-management level team. It was the director's hope as well as well as members of his team that each of the regional teams engaged with the Process Enneagram process would build their respective capacity to engage in ongoing learning and team development. In this way it was anticipated that a high level of synergy would emerge such that the goals and

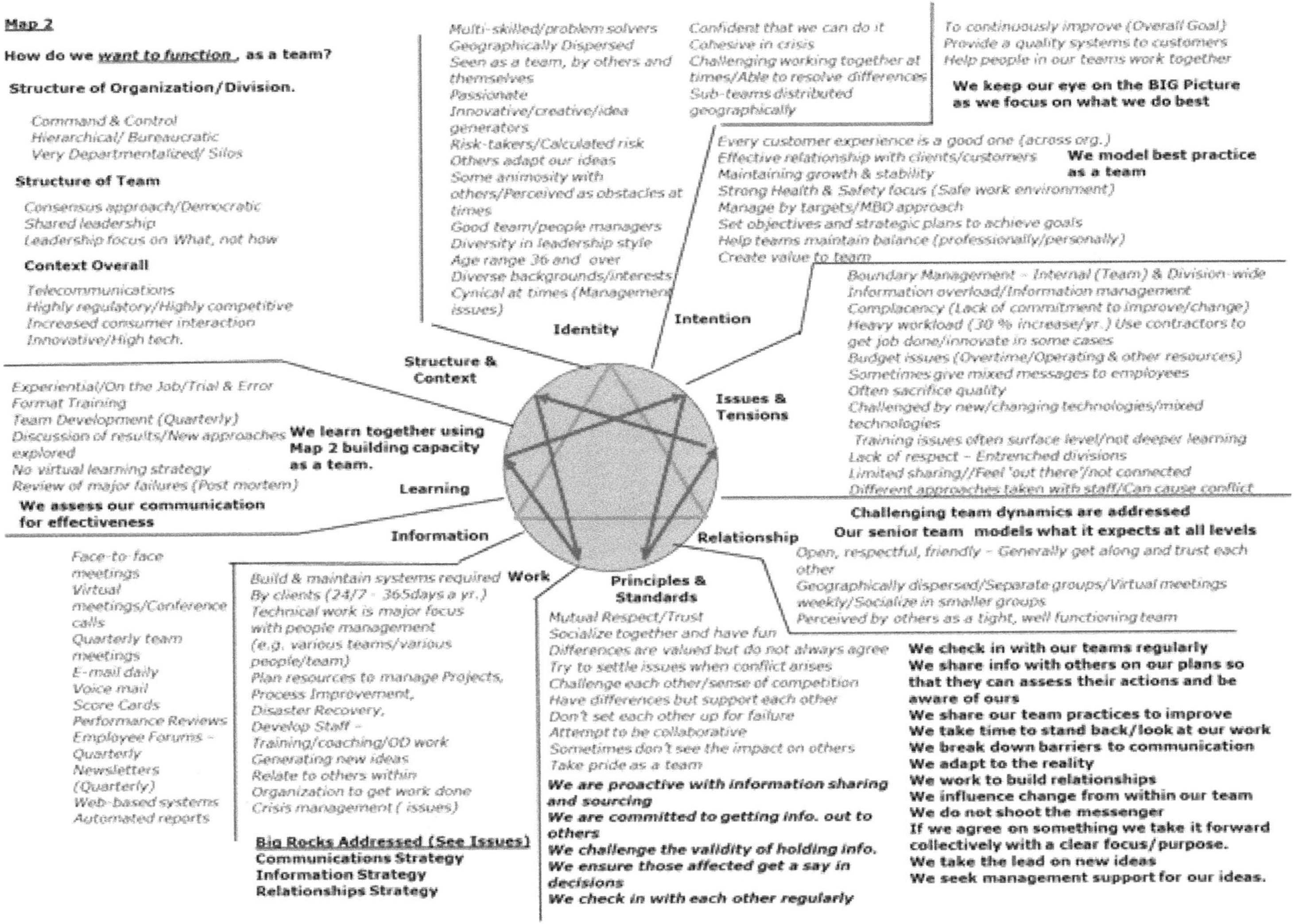

Map 2

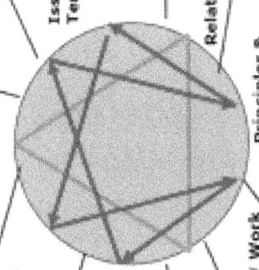

Figure 3 *Desirable Function as a Team*

Map 2

How do we want to function... as a team?

Structure of Organization/Division.
Command & Control
Hierarchical/ Bureaucratic
Very Departmentalized/ Silos

Structure of Team
Consensus approach/Democratic
Shared leadership
Leadership focus on What, not how

Context Overall
Telecommunications
Highly regulatory/Highly competitive
Increased consumer interaction
Innovative/High tech.

Experiential/On the Job/Trial & Error
Format Training
Team Development (Quarterly)
Discussion of results/New approaches explored
No virtual learning strategy
Review of major failures (Post mortem)
We assess our communication for effectiveness

Face-to-face meetings
Virtual meetings/Conference calls
Quarterly team meetings
E-mail daily
Voice mail
Score Cards
Performance Reviews
Employee Forums – Quarterly
Newsletters (Quarterly)
Web-based systems
Automated reports

Build & maintain systems required
By clients (24/7 - 365 days a yr.)
Technical work is major focus with people management
(e.g. various teams/various people/team)
Plan resources to manage Projects,
Process Improvement,
Disaster Recovery,
Develop Staff –
Training/coaching/OD work
Generating new ideas
Relate to others within
Organization to get work done
Crisis management (issues)

Big Rocks Addressed (See Issues)
Communications Strategy
Information Strategy
Relationships Strategy

We learn together using Map 2 building capacity as a team.

Multi-skilled/problem solvers
Geographically Dispersed
Seen as a team, by others and themselves
Passionate
Innovative/creative/idea generators
Risk-takers/ Calculated risk
Others adapt our ideas
Some animosity with others/Perceived as obstacles at times
Good team/people managers
Diversity in leadership style
Age range 36 and over
Diverse backgrounds/interests
Cynical at times (Management issues)

Confident that we can do it
Cohesive in crisis
Challenging working together at times/Able to resolve differences
Sub-teams distributed geographically

To continuously improve (Overall Goal)
Provide a quality systems to customers
Help people in our teams work together
We keep our eye on the BIG Picture as we focus on what we do best

Every customer experience is a good one (across org.)
Effective relationship with clients/customers
We model best practice as a team
Maintaining growth & stability
Strong Health & Safety focus (Safe work environment)
Manage by targets/MBO approach
Set objectives and strategic plans to achieve goals
Help teams maintain balance (professionally/personally)
Create value to team

Boundary Management – Internal (Team) & Division-wide
Information overload/Information management
Complacency (Lack of commitment to improve/Change)
Heavy workload (30 % increase/yr.) Use contractors to get job done/innovate in some cases
Budget issues (Overtime/Operating & other resources)
Sometimes give mixed messages to employees
Often sacrifice quality
Challenged by new/changing technologies/mixed technologies
Training issues often surface level/not deeper learning
Lack of respect – Entrenched divisions
Limited sharing/Feel 'out there'/not connected
Different approaches taken with staff/Can cause conflict

Challenging team dynamics are addressed
Our senior team models what it expects at all levels
Open, respectful, friendly – Generally get along and trust each other
Geographically dispersed/Separate groups/Virtual meetings weekly/Socialize in smaller groups
Perceived by others as a tight, well functioning team

Mutual Respect/Trust
Socialize together and have fun
Differences are valued but do not always agree
Try to settle issues when conflict arises
Challenge each other/sense of competition
Have differences but support each other
Don't set each other up for failure
Attempt to be collaborative
Sometimes don't see the impact on others
Take pride as a team
We are proactive with information sharing and sourcing
We are committed to getting info. out to others
We challenge the validity of holding info.
We ensure those affected get a say in decisions
We check in with each other regularly

We check in with our teams regularly
We share info with others on our plans so that they can assess their actions and be aware of ours
We share our team practices to improve
We take time to stand back/look at our work
We break down barriers to communication
We adapt to the reality
We work to build relationships
We influence change from within our team
We do not shoot the messenger
If we agree on something we take it forward collectively with a clear focus/purpose.
We take the lead on new ideas
We seek management support for our ideas.

Center labels: Intention · Issues & Tensions · Identity · Relationship · Structure & Context · Principles & Standards · Work · Learning · Information objectives established by the mid-level management team would permeate into and be reflected through the actions of the various regional teams. In effect, by working together using the Process Enneagram all teams would be working off the same 'song sheet' in achieving outcomes, even though each team would have unique and geographically sensitive ways of achieving these outcomes.

What occurred is the essence of this paper. What follows is an account of the outcomes achieved (or not) in this attempt to build a strong, cohesive and highly functional set of regionalized teams reporting to a mid-level management team who itself functioned and modeled the way in a similar manner.

OUTCOMES ASSOCIATED WITH TEAM DEVELOPMENT WORK

What follows is an account of the various outcomes resulting from the application of the Process Enneagram to team development work for a mid-level management team and its various geographically disperse regional teams. The manner by which this is presented reflects the various components of the Process Enneagram as outlined earlier in this paper.

The Influence Of Structure/Context On Effective Adoption Of The Team Goals, Principles, Relationships And Information Sharing Components Of The Process Enneagram

It was clear at the beginning of the team development work that the larger organization within which the mid-level management team operated was ruled from head office. They set the stage for how people (i.e., teams) were expected to work together, both in a directional sense and from a cultural perspective. This is represented in Maps 1, 2 and 3 as the 'corporate culture'. They specified KPI's (Key Performance Indicators) that were expected to be realized by all employees and monitored these regularly. Performance evaluations incorporated these KPIs as a significant factor in measuring individual and team success. Projects and project timelines were dictated from head office. Teams and the people with them were duty bound to work towards these measurements, sometimes at the expense of changing how the team managed itself, (e.g., relationships, information sharing, issues management, sense of identity, etc.) At times this would drive behaviors contrary to the principles and standards adopted by the various teams (e.g., managers would complain about each other due to the pressures of the job. One manager would want to get something done, while the other would resist cooperating because he had limited resources and might be left open to being compromised in his section of the operation). Various issues also stood in the way of living the agreed principles and standards of cooperating and supporting each other. For example, due to a lack of effectively scheduling resources someone working for one manager might commit another team to work overtime; the manager of that team would often have no say in this which would result in the experience of having someone else running his area/priorities. As a result information sharing would suffer since teams would quickly learn that providing information to others may adversely affect them and their performance as monitored by head office. A 'protection to survive' mode of operation became evident. Team members began to fear that if they went too far in the direction that the team development work proposed and the Director wanted they may be putting their necks on the line as corporate employees. The risk attached to this change was palpable demonstrating itself in a failure, at times, to fully commit to living the principles and standards agreed to as a team and in fully addressing the 'big rocks' agreed to by the team. The director and OD Consultant as well as several of the more committed team members were clearly able to see two systems opposing each other; one being the corporate system; the other the mid-level management team culture.

As those professionals working in the fields of systems thinking and complexity theory know, any change in behavior (i.e., difference way of holding oneself accountable to agreed principles and standards) that influences or effects a perceived or potential change in the larger system (e.g., head office) is most often perceived as a threat.

Culture can be very different at the team level compared with the corporate level. At the team level, in this instance, the director was willing and eager to try new approaches to team development, engage in learning new ways of functioning as a team, and not willing to simply follow the direction being given from above to produce 'outputs' measured as KPIs. Nor was he willing to simply accept the status quo of 'don't question/just do' and be rewarded for that behavior. He visualized a different way of working as a team.

What was anticipated by the director in undertaking this team development work using the Process Enneagram approach was that just because you are trying to create a better team culture, that shouldn't prevent you from reaching the targets of the organization, whatever that organizational culture might be. If you are successful at achieving the targeted objectives it should be a win-win situation. What was observed, however, is that this did not happen in the situation presented here. The question is why.

Perhaps something did happen in the minds and hearts of team members who experienced the Process Enneagram process and whatever that is lies somewhat dormant in the residual team members, since many of the original members including the director and several managers have since left. We don't know the answer to this question but it suggests that culture and structure play a dominance role on influencing the degree and level of commitment that can be achieved in establishing principles and standards that might direct a team or team members to behave 'differently' than what is expected.

The Didactic Nature And Process Of Getting To An Agreed Set Of Principles/Values

Over several years working with the mid-level management team and its various regional teams, a number of iterations of the Process Enneagram mapping exercise were employed. Initially the team development work moved quite expeditiously from Map 1 (Figure 2) to Map 2 (Figure 3)(referencing the work of the mid-level management team), particularly in defining key principles and standards that the team felt needed to be established. In addition, it seemed at the start that clarifying intentions and issues, defining the 'big rocks' of work to be done and engaging in meaningful dialogue was fairly easy. While there was a degree of initial reticence about engaging in the Process Enneagram process, perhaps perceiving it as somewhat touchy-feely, team members seemed more accepting of the value of this work as it proceeded and in how it related with other activities. There was a strong growing sense of commitment towards more open dialogue in addressing team development issues. People were different after engaging in the process in relation to their interpersonal associations and openness to talking about what they were doing together and how they wanted to be together. It seemed that the initial process 'validated' what team members were thinking and wanting. There was a safe 'container' created through this process that helped to support a genuine exchange of ideas within an effective environment for dialogue.

Subsequent team development sessions in year two at the mid-level management team level, took the initial work and tried to improve it in areas such as 'the big rocks' and in developing Map 3 (Figure 4) which attempted to build on how the team could be even better with regard to living the principles and standards it had espoused and in using the nine components of the Process Enneagram to assist them in their team development work.

Map 3

How do we want to function, as a team?

Structure of Organization/Division.

Command & Control
Hierarchical/ Bureaucratic
Very Departmentalized/ Silos

Structure of Team

Consensus approach/Democratic
Shared leadership
Leadership focus on What, not how

Context Overall

Telecommunications
Highly regulatory/Highly competitive
Increased consumer interaction
Innovative/High tech.

Identity

Multi- skilled/problem solvers
Geographically Dispersed
Seen as a team, by others and themselves
Passionate
Innovative/creative/idea generators
Risk- takers/Calculated risk
Others adapt our ideas
Some animosity with others/Perceived as obstacles at times
Good team/people managers
Diversity in leadership style
Age range 36 and over
Diverse backgrounds/interests
Cynical at times (Management issues.)

Intention

Confident that we can do it
Cohesive in crisis
Challenging working together at times/Able to resolve differences
Sub-teams distributed geographically

To continuously improve (Overall Goal)
Provide a quality systems to support
Help people in our teams work together
We keep our eye on the BIG Picture as we focus on what, we do best

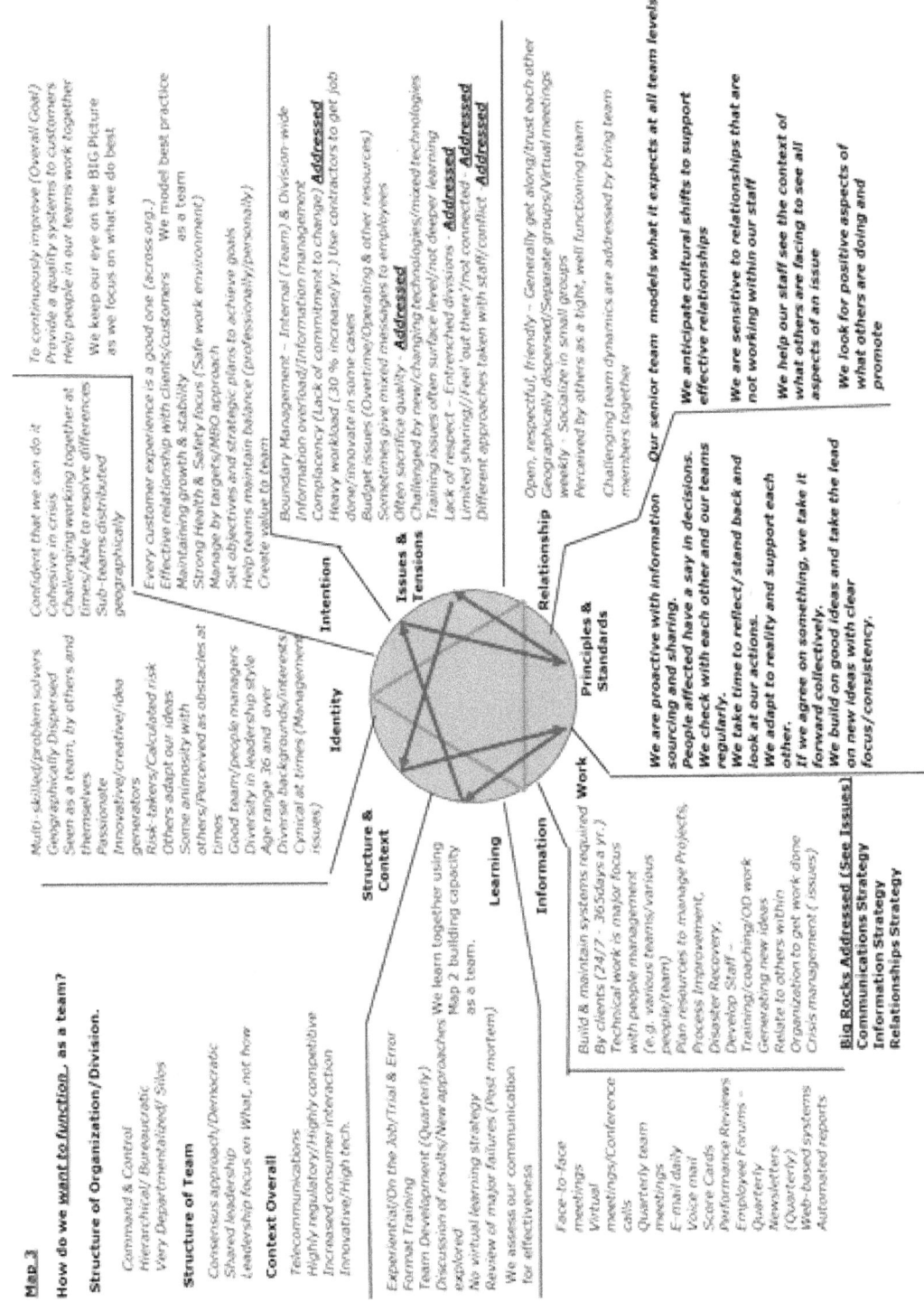

Issues & Tensions

Boundary Management – Internal (Team) & Division wide
Information overload/Information management
Complacency (Lack of commitment to change) **Addressed**
Heavy workload (30 % increase/yr.) Use contractors to get job done/innovate in some cases
Budget issues (Overtime/Operating & other resources)
Often sacrifice quality - **Addressed**
Challenged by new/changing technologies/mixed technologies
Training issues often surface level/not deeper learning
Lack of respect – Entrenched divisions - **Addressed**
Limited sharing//Feel 'out there'/not connected - **Addressed**
Different approaches taken with staff/conflict **Addressed**

Every customer experience is a good one (across org.)
Effective relationship with clients/customers   We model best practice
Maintaining growth & stability   as a team
Strong Health & Safety focus (Safe work environment)
Manage by targets/MBO approach
Set objectives and strategic plans to achieve goals
Help teams maintain balance (professionally/personally)
Create value to team

Relationship

Open, respectful, friendly - Generally get along/trust each other
Geographically dispersed/Separate groups/Virtual meetings weekly - Socialize in small groups
Perceived by others as a tight, well functioning team
Challenging team dynamics are addressed by bring team members together

Our senior team models what it expects at all team levels

We anticipate cultural shifts to support effective relationships

We are sensitive to relationships that are not working within our staff

We help our staff see the context of what others are facing to see all aspects of an issue

We look for positive aspects of what others are doing and promote

Principles & Standards

We are proactive with information sourcing and sharing.
People affected have a say in decisions.
We check with each other and our teams regularly.
We take time to reflect/stand back and look at our actions.
We adapt to reality and support each other.
If we agree on something, we take it forward collectively.
We build on good ideas and take the lead on new ideas with clear focus/consistency.

Structure & Context

Learning

Experiential/On the Job/Trial & Error
Format Training
Team Development (Quarterly)
Discussion of results/New approaches explored
No virtual learning strategy
Review of major failures (Post mortem)

We learn together using Map 2 building capacity as a team.

Information

We assess our communication for effectiveness

Face- to- face meetings
Virtual meetings/Conference calls
Quarterly team meetings
E- mail daily
Voice mail
Score Cards
Performance Reviews
Employee Forums
Quarterly Newsletters (Quarterly)
Web- based systems
Automated reports

Work

Build & maintain systems required
By clients (24/7 - 365 days a yr.)
Technical work is major focus
with people management
(e.g. various teams/various people/Team)
Plan resources to manage Projects,
Process Improvement,
Disaster Recovery,
Develop Staff,
Training/coaching/OD work
Generating new ideas
Relate to others within
Organization to get work done
Crisis management (issues.)

Big Rocks Addressed (See Issues)
Communications Strategy
Information Strategy
Relationships Strategy

Figure 4 *'Revised' Desirable Function as a Team*

Relationships both among team members and among other teams became a key aspect of discussion. More sensitivity was given to relationships that were not working and in helping other regional team members see what they are doing and why. Effective relationships and relationship-building were seen as essential to overall performance and success as a team and as an operational unit.

Acceptance of the work to be done in shifting team culture to a higher level of performance was evidenced in the early stages of this work and to some degree throughout the process which led towards the engagement of the various regional teams in this process. It was also recognized that building and maintaining effective team relationships was a challenge because of how wide-spread the operational unit was across a vast geographical area. The commitment that the mid-level management team members (i.e., managers) displayed to seeing to the work of team building across the operational unit was a bright and shared moment of success in this work. After nearly two years of working to improve team functioning both at the mid-level management team and regional team levels it was generally felt that the application of the Process Enneagram to team development had been effective.

As Map 3 (Figure 4) demonstrates, the mid-level management team took the job of building an effective high functioning team seriously at the beginning of year three. Key issues were addressed, relationships were strengthened and principles and standards were abbreviated to the essential elements of what each team member took as their guidelines for performance as team members and as a team.

However, when the mid-level management team attempted to push further in year three, exploring deeper ways of understanding themselves as a team with the application of different tools and approaches (e.g., spiral dynamics work, e.g., Beck & Cowan, 1996), it seemed to become more difficult for members of the mid-level management team to grasp how this information, while interesting to some team members, could be used. Perhaps this was too much of a leap at that time in adding new concepts and pushing ahead at the expense of further grounding the work undertaken to date. This issue is reviewed later in the paper.

As a result, the team development effort lost some momentum in year three both at this level and at the regional team level. What appears to have occurred is that each of the mid-level management team members had a different 'capacity' to understand, accept and employ the work of building a stronger team based on the application of the Process Enneagram at a deeper level and this transmitted to and impacted the work at the respective regional team level. What began to be evidenced at this stage in the process was a tacit commitment to addressing the 'big rocks' but a lack of follow-up and commitment to actually doing the work, with the subsequent effect of seeing in many instances these big rocks fall through the cracks? While there were verbally expressed great intentions about getting on with these commitments, little was actually done. It felt like a passive-aggressive behavior at times. While some positive actions were evidenced (e.g., communications efforts and information sharing were seen as team members arranged to connect on important issues) this was far short of expectations.

Taking the big rocks from concept to action seemed to be a struggle. Similar to Edward Debono's (1985) technique, the various teams were excellent at coming up with ideas but not good at implementation. It seemed that team members at all levels within the operation displayed various levels of commitment regarding the question of "What's in it for me" vs. "What's best for the team". The good work associated in the early stages with the Process Enneagram began to dissipate into a so-so acceptance of a team culture that this was not really going to work effectively within the corporate structure. What we are left to assume here is that both the added weight in year three of deepening the Process Enneagram work (e.g., application of Spiral Dynamics, Beck & Cowan, 1996) and the constant impact of the corporate culture to keep things as they were eventually lead to a lessening of commitment and effort on the part of various team members.

The Rationalization Of Intention And Work In Achieving Effective Work Performance And Its Effect On Other Components

One of the most essential factors for team success is the ability to link intention (i.e., purpose) with the work that gets done. So often teams set goals and objectives that are well established in theory but not well executed in practice. Often work gets done that is not essential or relevant to the intentions for which the team is established. Work gets done, but does not lead to completion of goals or objectives. One way to measure this dynamic is through the Process Enneagram process. As can be seen in the earlier maps, particularly as we look at Maps 2 and 3 (Figures 3 and 4), i.e., areas where the team improved in clarifying intention and adjusting the work accordingly, intention and work are much more closely aligned. This was a positive outcome of the overall team development process. Team members could readily see that by achieving clarity and defining work more clearly both their time and effort in achieving performance measures were affected. Clearly this had much to do with how they shared information and how they build effective relationships based on established principles and standards.

Dealing With Tensions/Issues In An Integral Manner Through The Process Enneagram And Seeing The Effect On And/Or Contribution Of Other Components

As stated at the beginning of this paper, each of the nine (9) components of the Process Enneagram are connected to every other component in an integral, integrated and emergent manner. As one component changes so do all components. This is important to remember as teams begin to change and adjust their sense of identity, build more effective relationships and share information in a more direct and supportive manner. Perhaps the best evidence of how these components inform and influence each other can be seen in the changes that occur over time with issues and tensions. As Map 1 (Figure 2) displayed there were a considerable number of issuers and tensions evident at the beginning of the team development work, most notably; complacency and lack of commitment, giving mixed messages, sacrificing quality, showing a lack of respect, inconsistencies in dealing with issues, limited sharing of information, etc. As the mid-level management team continued to work with the Process Enneagram and address the way they functioned as a team these issues became less evident. In Map 3 (Figure 4) we see a much more proactive approach to addressing these issues and others evidenced by the 'big rocks' that are listed under the team's work component. As well, the degree of information sharing, relationship-building and learning all show a marked increase. Key principles and standards are clarified and 'lived', not simply espoused. There is a palpable level of accountability for how the team operates which is not evident in earlier stages of the team development work.

Similarly, as relationships improved, information sharing increased and a greater sense of team identity emerged, issues become more manageable, talked through and shared. While new issues emerged over time they seemed to be handled in a more effective manner without becoming obstacles for the team.

In many respects the functioning of a team resembles the operation of any organism. If for example an animal like a dog has a burr in its foot all attention is focused on it often creating further injury and discomfort as the dog attempts to remove the burr. Not only does the animal fret and become anxious in this situation, its mood and manner can become aggressive and mean. So too do teams experience the burrs of issues and tensions that lay in their path. Proper attention to the pathways taken, the support offered and to proactive efforts to address issues arising can mean the difference between a well-functioning team and a dis-functional one. Agreed upon principles and standards act as a container for holding the team accountable to itself and in helping it measure its performance as a team when dealing with difficult issues and tensions that arise.

It is evident in this work that while each of the nine components of the Process Enneagram can and do change, often in emergent and unpredictable ways, the connection among these components helps to bring awareness and opportunity for change and adjusts one's behavior towards being a highly functioning team. In this way the Process Enneagram serves as an accountability framework to hold the team accountable to itself.

KEY SUCCESS FACTOR—PRINCIPLES AND STANDARDS

Perhaps of all of the components contained within the Process Enneagram the one that is most critical is principles and standards. It has been said that when one is in the desert with nothing but sand on every side and with no visible point of reference, principles and standards act like a stick planted in the sand. By reference to the stick one is immediately able to judge the distance from the stick as one moves forward. Teams use principles and standards as guidelines, reference points and benchmarks to judge how they are functioning as a team. Principles and standards represent the inherent and explicit set of values one holds and as stated earlier, provides an accountability framework for team behavior and action. When all else fails the team and its members can refer to these principles and standards as a way to open the dialogue and have meaningful conversations. Glenda Eoyang[4] has stated this quite clearly in referring to, "differences that make a difference".

In discussions with the director of the mid-level management team, building on the principles and standards adopted were paramount to changing the culture of the team. A coaching wheel was developed to effectively gage how each team member, including the director, was living the principles and standards. Each principle and standard was inserted in one of the wheel segments/spokes and rated on a scale from 1-10 to determine compliance and to identify how each team member and the team as a whole was living the principles and standards. What occurred in using this process was an indication that continued use of the coaching wheel could provide a good indicator of growth and development in the team's ability to live the principles and standards. While the team remained together, prior to the director leaving the organization after year three of this work, this capacity to monitor and adapt team behavior based on the team's principles and standards proved very effective.

---

4. Refer to Glenda Eoyang's work—Glenda Eoyang, Ph.D. is a pioneer in the field of complexity and human systems dynamics (HSD). Glenda is a gifted teacher and practitioner who blends remarkable theoretical insights with extraordinary practical experience to help others observe, decide, and act in the most complex and challenging situations. She leads a network of 135 certified Human Systems Dynamics Professionals who use HSD theory, models, methods, and tools to engage creatively with individuals, groups, and institutions.

One further adaptation of the use of tools to monitor performance was a behavioural grid[5]. By showing what should be reinforced and what should be eliminated; what should be paid attention to and what should be done over and over, the team was able to adapt and build its capacity as a highly functioning team.

In essence, while some might say are the 'soft' aspects of team work, principles and standards are in fact the 'hard' ones. By holding team members accountable to their espoused principles and standards and not letting them escape this accountability, teams improve the way the function. It's as simple as that. Without these guidelines and set of espoused values teams wander and flounder in a sea of indecision, strife and poor performance (e.g., Heifetz, 1994).

OBSERVATIONS FOR FUTURE WORK

While one can learn from doing this work and in further applying that learning with hopes of improvement in team functioning, organizational contexts and new team members change the dynamics of team development work and present challenges. Furthermore, as one works towards improvement and attempts to address these challenges further issues arise related to motivation, commitment and risk. One example of this is the lack of tolerance that can set in when teams and team members loose the capacity or interest for trying new things. In the team development work discussed in this paper the mid-level management team was asked at one point to try a deeper level of learning about each other's value structures using an approach called Spiral Dynamics (Beck & Cowen, 1996). Some of the team members did not have the capacity to go deeper. However, these people were a good fit for the existing corporate culture. Here, as was mentioned earlier in this paper, we observe systems competing with each other; the more formative thinking team members who engaged in new learning and the more rational/logical members who had engaged enough after the first two to three years and did not see the value in doing more. The lesson from this perhaps is that teams have their limits and at some point team leaders need to recognize these limits and work to strengthen what has already been accomplished without pushing further. In the example presented here, the team leader (i.e., director) wanted to develop the team members and take them to a deeper level of knowing and being. People were encouraged to think not only about how they work as a team but about how they function as people within a team. There was a desire to help them reflect on their personal learning so that they could be even better as individuals and as team mem-

---

5. The use of a behavior grid reported here is founded on the work of Richard Knowles (see earlier references) whereby a model showing on one axis the things that are important to do and on another axis the things that should be reinforced, teams can self-measure and adapt their behavior.

bers. This was to many a gift that came out of the Process Enneagram work providing some people with a renewed and expanded self-perspective. For others, this was not so.

The lasting value and legacy of this work, based on selected discussions with team members and the director and based on the authors' own observations, is that there currently remains a better communication and improved level of team performance evidenced within the mid-level management team compared with other teams in the organization. Recent corporate surveys have been undertaken within the company since the Process Enneagram work was completed some years ago and in measuring levels of engagement have shown that engagement has improved as has loyalty and relationships among team members. This is positive news for those wishing to use the Process Enneagram to improve these components of team performance.

A further observation made in preparing this paper based on discussions with selected team members is that a substantive level of commitment and acceptance by team members and members of the regional teams towards the Process Enneagram process has been established The metrics on this work based on anecdotal evidence from selected team members suggested that while there were indications of team improvement in communications, loyalty and relationships, the team members sampled generally felt that improvement related to the corporate metrics (e.g., KPI's) which were more about each person meeting output objectives was not significantly affected and in several areas improved upon. Unfortunately at the corporate level there is no established set of metrics for measuring team functioning itself. This remains an outstanding issue for teams as they strive to work towards improved team functioning.

Finally, as stated earlier, as the team development work went deeper (for example, working on the 'big rocks' and exploring the work on Spiral Dynamics), many didn't want to put the energy into this work. The sense was that balance in life takes over. Questions that might be asked include, "Why put in the extra effort when there is no clear reward corporately", or "When the sacrifice for improvement means letting other personal goals suffer, why do so". It appeared in this work that the younger and less encumbered team members showed a higher level of excitement about the learning. Perhaps there are certain periods in one's life when this level of commitment to learning and improvement is important and other times when this wains. One further question that might be posed here is, "Would the Process Enneagram work be more effective when the team is new and younger? Further, "Would it work better when there is more to gain than lose?" As life progresses we measure risk differently. The overall success of this work is dependent on how much people are prepared to learn and risk improvement in themselves as well as the team. No risk; no reward.

CONCLUDING REMARKS

Throughout this paper, the authors have attempted to show how the application of the Process Enneagram can in many ways assist in team development efforts in bringing clarity and accountability to this work. Most importantly, it is a testament to the value of helping teams function more effectively by providing a safe container for dialogue and a framework for improved functioning. By using this complex systems 'tool' to engage with complex, adaptive systems behavior evidenced in teams and team members, team development work becomes a far more generative and emergent process. Many of the approaches used in team development lack this finesse and organic capacity to adapt with the team, to be fluid and open to change and to be reflective of the emergent and unpredictable nature of team development work. The Process Enneagram allows us to move from reductionist and mechanistic ways of monitoring and measuring team functioning to a more human sensitive and adaptive approach which recognizes that human beings function more as complex, adaptive systems than as machines. Team development work is not about 'fixing' people; it's about helping people adapt and adjust to ways of working together that bring meaningful outcomes. The team development case study that has been presented and examined in this paper reflects work that continues to be undertaken using the Process Enneagram. It is the authors' hope that this account will assist others in undertaking this work and give further support to the use of a complex systems, self-organizing approach to team development.

REFERENCES

Knowles, R.N. (2002), *The Leadership Dance: Pathways to Extraordinary Organizational Effectiveness*, ISBN 0972120408.

Beck, D. and Cowan, C. (1996). *Spiral Dynamics: Mastering Values, Leadership and Change*, ISBN 1557869405.

de Bono, E. (1985). *Six Thinking Hats: An Essential Approach to Business Management*, ISBN 0316177911.

Heifetz, R.A. (1994). Leadership without Easy Answers, ISBN 0674518586, with reference to the notion of 'adaptive challenge'.

Chapter 8

A TAO TRANSFORMATION LEADERSHIP MODEL FOR THE PROCESS ENNEAGRAM©

Richard Bergeon & Caroline Fu

The Tao transformation leadership Model, developed from a rediscovered leadership concept conceived five-thousand years ago, draws upon modern/classical physics, ancient Chinese cosmological science about Nature's phenomena, and practical experience. This model illustrates a correlation between the nine perspectives used in The Process Enneagram with nine states of energy-flow that emerge during complex transformations cycles. In this model, energy-flow represents an abstraction of energy presence and movement in a spatiotemporal field of relationships. The energy-flow transformation concept offers a dynamic theoretical and philosophical underpinning to support and explicate further the implicate order in the Process Enneagram. This paper provides explanations on how the perspectives hold attributes of energy and how to provoke critical thinking to transform and balance energy engaging the collective in purposeful interventions leading to success.

A TAO TRANSFORMATION LEADERSHIP MODEL FOR THE PROCESS ENNEAGRAM

The human mind is ever watchful for patterns. Our greatest scientific achievements come from recognizing patterns. Many managers search for patterns that might assist in explicating complexity in the world. As operational managers, our own attention turned to process tools that worked with patterns as we struggled with complexity. We believed there must be tools to help reveal paths to success hidden in the chaos of life.

In our search, we discovered an ancient Chinese pattern (☯) called Tai-Ji (supreme ultimate), representing the basic Nature's day-night phenomenon (Fu, 2008). Nature, and its cycles, inspired the Book of Changes (aka I-Ching) that pre-dated Tao philosophy and the Book of Tao (Tao-Te-Ching). Capra (1975/1991) informed us that ancient Eastern sages held time and change to be the essential features of a dynamic world (p. 24). Their understanding of science did not require everything to fit into perfect straight lines or shapes (p. 163). Unlike the early Greek concept of the world being a manifestation of multiple opposites under tension (p. 20), Eastern sages saw the world as interrelationships in which opposites were complementarities (p. 24) rather than dichotomies. For them, knowledge that we talk about as rational is only "relative," and abstraction a "crucial feature of this knowledge" (p. 27).

We were drawn to the ancient pattern of change (Tai-Ji) now commonly called Tao. Tao represents a pattern in cyclic motion within which two interactive polar extremes appear as opposites; yin and yang (Fu, 2008). When the yang reaches its climax, it retreats in favor of yin; and when yin climaxes it retreats in favor of yang (Capra, 1975/1991: 106). The movements of the Tao are natural and it is only necessary to move with the flow of the Tao in order to sustain harmony, called "wu-wei" (p. 107). Vaill (1989) saw wu-wei in "the Western managerial idea of 'taking action' [yang]" and "nonaction [yin]" (p. 177). Siu (1978) relabeled the complementarity of "acting" and "nonacting" as "quasi-acting," (p. 83).

Over the course of decades, we worked to further our understanding of the ancient Tao concept of yin-yang dynamics. While attempting to validate our thinking, we came across Richard Knowles's (2002), The Leadership Dance and found parallels in his process work. In the early half of the twentieth century George I. Gurdjieff saw a set of patterns in life that he named the enneagram (Knowles, 2002: 107). Knowles applied Gurdjieff's energy flow in the enneagram to process in much the same way that we were applying the flow in the Tao.

In this chapter, we relate a basic understanding of the Tao, explore how energy-flow in process parallels Knowles' Process-Enneagram and makes contributions to understanding decision making strategies. Our thoughts about the energies that exist in Tao presence interacting with the eight I-Ching "trigrams" (Fu & Bergeon, 2011: 18) reveal some new perspectives regarding the use of the Process Enneagram tool.

The Tao

Five thousand years ago, Eastern sages formed a concept of being and time that is somewhat unfamiliar to the Western mind. For the sages, being was inextricably linked with change. Nature provided endless references that were abstractions of the forms that change adopted. Change was perceived as cycles, transformations, and transmutations. In Tao, either-or was dismissed and becoming-adjusting took precedence.

Figure 1 *The Tao*

The symbol of the Tao (an artistic version in Figure 1) is widely recognized but not well understood. The swirl shape of darkness and lightness implies motion and enfolding (Bohm, 1980) like the cycle of day/night and the change of the seasons. The opposite color dots in each is a nod to the fact that never is one aspect devoid of the other—there is shadow in the day, light in the darkest night. Tao's message is that being exists in change. Identity, therefore, associates with change. Both, change as process and identity as manifestation of being, are the locus of existence.

In the phenomena of Nature, the ancient sages found four sets of complementarity that provided abstractions of Nature's various degrees of yin/yang polarity energies manifested as: Earth/Heaven, Wind/Thunder, Fire/River, and Mist/Mountain. In the Inner-World Arrangement they are arranged in the octagonal vertexes as shown in Figure 2. An enneagram arrangement is presented in Figure 3 below.

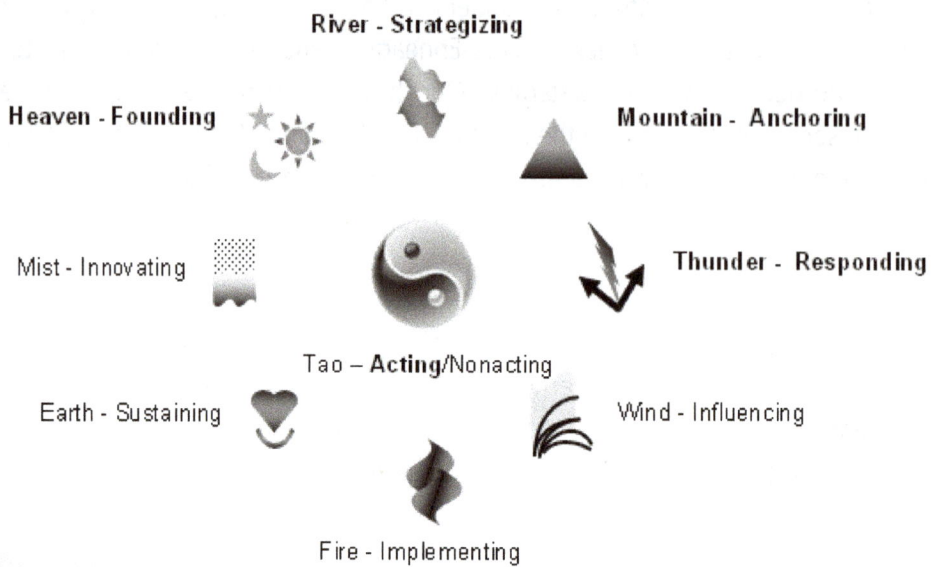

Figure 2 *The Tao Model arranged in the octagonal Inner-World Arrangement (Fu, 2008)*

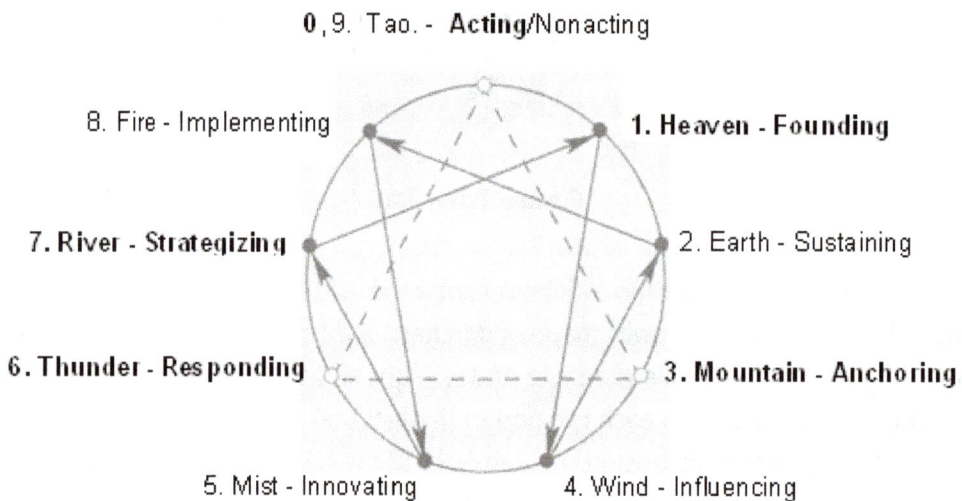

Figure 3 *Tao Model as Enneagram*

We use physics nomenclature to identify the yin/yang as potential/kinetic energies. In both Figures 2 and 3, the four energies indicated in bold text are active (yang) or kinetic energies while the other four, in non-bold text, are passive (yin) or potential. Each complementarity has degrees of yang/yin, active/passive, and forceful/waiting aspects. Note that wu-wei harmony is embraced by all. In the Tao/enneagram comparison, [0] is acting, and [9], nonacting, which is often experienced as waiting.

Nature	Energy	Translations	Process	Focus
Heaven	Founding	Resolute, Firm, Positive, Advance, Protect	Resolve	Intention, Vision, Goals, Purpose, Reason for Being, Power, Authority
Earth	Sustaining	Adapt, Accept, Frugal, Nurture, Substantive	Resource Acquisition/ Allocation	Plant/equipment, Employees/members, Wealth, Assets, Goodwill, Loyalty of employees and customers, Others' awareness and respect
Mountain	Anchoring	Accountable, Trustworthy, Reserved, Dependable, Faithful, Committed	Context Maintenance	Values, Beliefs, Traditions, History, Experience, Systems, Policies, Processes, Procedures, Leadership Structure
Wind	Influencing	Influence, Charismatic, Aware, Purposeful, Disperse	Climate Maintenance	Morale, Expectations, Wants & Needs, Obligations & Commitments, Prestige
Mist	Innovating	Innovative, Agile, Intuitive, Playful	Exigency	Idea, Inspiration, Insight, Revelation, Competition, Shifts, Opportunity
Thunder	Responding	Thoughtful, Reflective, Swift, Impulsive, Communicative	Communication	Interactions, Verbal/ Nonverbal Responses, Reactions, Feedback
River	Strategizing	Judging, Courageous, Competitive, Confident	Implementation	Strategies, Plans, Tactics, Projects, Incentives, Disincentives
Fire	Implementing	Illuminative, Responsive, Dependent, Energetic	Progress	Anticipated & Unanticipated Outcomes, Achievements, Challenges, Accidents, Crises
Tao	Being/Doing		Reflection	Nonacting/Acting

Table 1 *Tao and the Energy-flow in Being/Doing*

A search of Chinese texts, in both Chinese iconography and many English translations, yielded information about the human energy the Nature-based abstractions reference. Ideas emerged about the people's being and doing, the energy-flow and action process on a field where change played out. The texts yielded the attribute descriptions listed in Table 1.

The Living Systems Paradigm

Each of the energies are constantly interacting much as in the manner that all organizations are composed of individuals who interact with each other in complex ways. As they carry out the satisfaction of personal wants and needs they form complex structures of individual, small group, and large group actions. People self-organize around shared interests, designated responsibilities, and formal and informal authorities. These activities tend to form a pattern that is replicated.

Knowles (2002) presented what he called "The Living Pattern" (pp. 33-35). In the Living Pattern we find a remarkable set of coincidences that indicates a correspondence between the Tao Model and Process Enneagram. As seen in Table 2, both Knowles' (2002: 39) Living Pattern and Rhodes' (2010: 83) fit process actions to the enneagram that coincide with the energies seen in the Tao energy-flow. For example, the first step [1] in the Living Pattern is becoming clear about the organization's intentions. This aligns with the Tao of Founding in which intention, vision, goals, etc. are identified. In step [2], agreement is reached on how the participants will interact. The leadership now exhibits the trait of Influencing as everyone's expectations, wants, and needs are taken into account and external commitments and obligations are factored into Intention. The organization works through the process flow arriving at a conclusion in which their experiences become part of the Living Pattern.

Knowles (2002) saw self-organization (p. 33) as expressed in Figure 4. The energy is all active/kinetic. Identity is the product of both Information and Relationships [0 and 3 and 6]. The Tao model similarly shows us that Resolve is constantly adapted by an unbroken flow of Communications of experiences that forewarn against crisis and that leads to Context Maintenance producing shared values, changes in policies, and internal systems. Knowles provided that steps [1, 4, and 5] in the Living Pattern remain open and repetitive leading to progressive improvements and adaptations.

Living Pattern Steps	Knowles's Action	Rhodes's Action	Tao Model Energy-flow
1. We talk together getting really clear.	Intention	Defining	Founding: Resolve intention, vision, goals, purpose, reason for being, power, authority
2. We agree on how to play the game, and practice.	Principles & Standards	Emotional Commitment	Influencing; Climate maintenance affects morale, expectations, wants & needs, obligations & commitments, prestige
3. We dialogue on these in the light of 1 & 2	Issues	Defining	Sustaining: Resource acquisition and allocation—plant/equipment, employees/members, wealth, assets, goodwill, loyalty of employees and customers
4. These develop as we do 1-3	Relationships	Organization	Anchoring: Context maintenance—values, beliefs, traditions, history, experience, systems, policies, processes, procedures, leadership structure
5. Open and accessible to all	Information	Following through	Responding: Communication interactions spawn verbal/ nonverbal responses, reactions, feedback
6. People and self-organize	Structures and Context	Completion	Implementing: Progress yields anticipated & unanticipated outcomes, achievements, challenges, accidents, crises
7. This gets done better & better	Work	Problem-solving	Innovating: Exigency produces ideas, inspirations, insights, revelations, inventions, transforms, opportunities
8. We keep finding better ways.	Learning	Opening up	Strategizing: Implementation brings forth adaptations, plans, tactics, new projects, incentives, disincentives, competition

Table 2 *The Living Pattern Extended by Tao Energy-flow*

The Command And Control Paradigm

In Nature balance prevails. Kinetic energy that is unused is stored and shifts to potential energy. This also happens in organizations. Knowles (2002) highlighted what he called the Command and Control Paradigm reproduced in Figure 5 (pp. 31-32). The desire for predictability and control motivates leaders to act in ways that lead to unintended consequences. Vagueness about intentions and organization directions lead to getting stuck on issues [0 to 2]. What happens is the process flow in which assets and employees are misallocated, loyalty is squandered. Issues dictate a solution [2 to 8]. Unanticipated outcomes create challenges and crises. Leaders solve

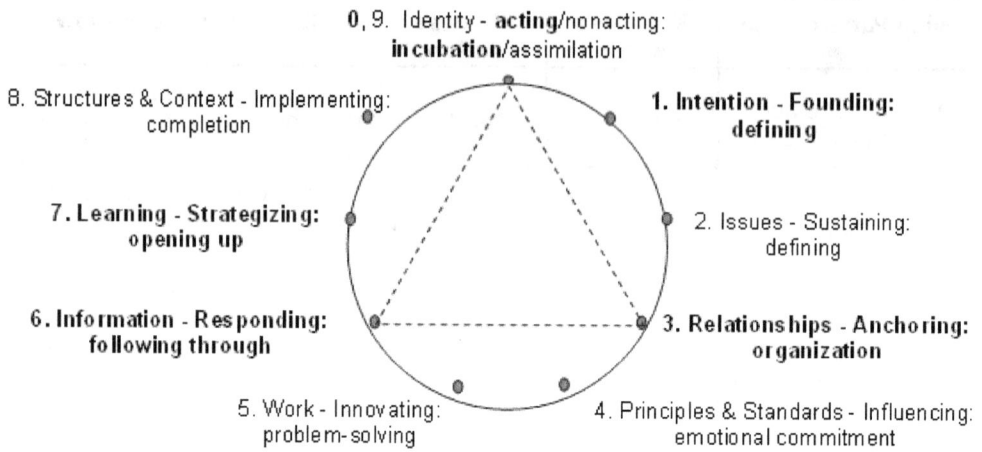

Figure 4 *Self-organization*

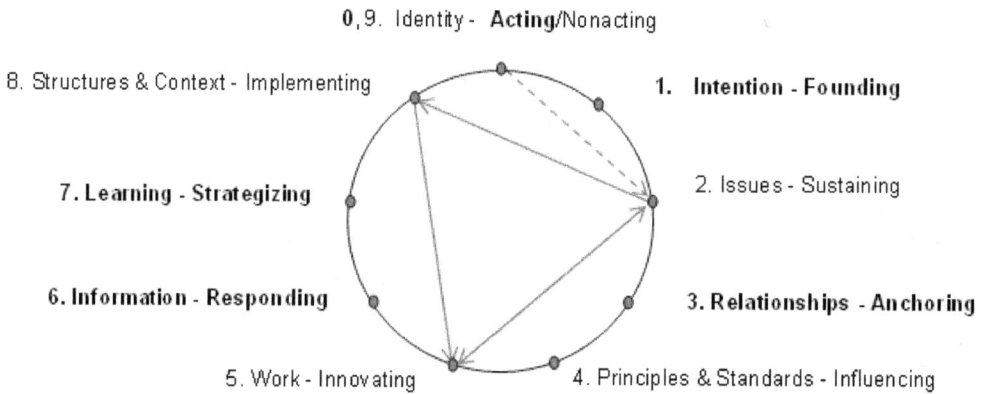

Figure 5 *Command & Control Paradigm*

the immediate problem looking to opportunities, imitating competitors, etc. imposing an answer [8 to 5]. Insights and revelations inspire arguments over resource allocations and asset distributions. The result is often repeated cycles of waste leading to unsustainable losses (bouncing between [2 and 5] until identity and intentions are worked out.

From the potential/kinetic energy perspective, Knowles's (2002) Command and Control Paradigm is out of balance since all the energy is yin/potential. The initial energy in Identity is unclear so the actions are all non-directive—waiting. This causes issues that are not coped with adequately so energy builds up: seen as frustration while waiting for clarity. Imposing an answer results in Work but the output does not resolve any issues (p. 32).

Figure 6 *Tao Model of Knowles's (2002) Command & Control*

In the Tao model (Figure 6) a similar pattern arises but the interaction of Knowles's (2002) [9, 2, 8, 5, 8]; then again, [9, 2, 8, 5, 8 . . .] becoming a long cycle. Whatever changes are made, resources will continue to be used unproductively. Resource exhaustion issues eventually arise and only when the potential energy builds up to become a crisis will the cycle be broken. Until then, ideas about what to do are ineffective and the organization waits to see what effects decisions will produce. Unintended consequences lead to benign actions until the issues are ultimately resolved by importing energy from outside the process flow.

The Tao Model And Transformation

Knowles (2002) provides other examples of broken systems in which over-reliance leads to broken response systems. Two of these are Operational Management, in which leaders become focused on things, and Strategic Leadership, where leaders become fixated on ideas. In both examples a response creates issues because energies are out of balance and the system must rebalance, but is unable to harmonize without help external to the operational or strategic process. Figure 7 shows the Strategic Leadership diagram (p. 41).

Knowles (2002) wrote that constant attention to Intention [1] increases focus on the needs and wants of customers, clients, and the community driving new plans, strategies, and incentives creating a secondary loop. In the primary loop, Learning [7] simultaneously drives changes to Principles & Standards [4] and Intention [1], but In-

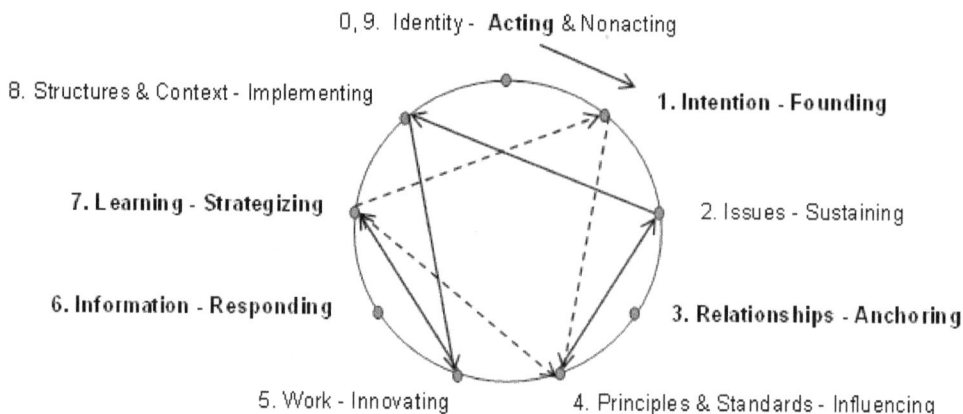

Figure 7 *Knowles's (2002) Strategic Leadership Overuse*

tention [1] also drives changes to Principles & Standards [4]. There is too much kinetic energy and insufficient potential energy. The organization initially finds itself acting too frequently and not being reflective about what it is doing or what the future may hold. This pressure for change to Principles & Standards [4] creates Issues [2] followed by a series of potential energy activities [2, 8, 5] that put energy out of balance again until Learning [7] restores the kinetic energy cycle in an attempt to rebalance: the organization energy shifts from kinetic, to potential, to kinetic, and may continue to repeat.

Imported Energy Influences And The Process Enneagram

Katz and Kahn (1966) observed that organizations are open systems possessing built-in goals subject to external influences, and it should be possible to understand how the pursuit of those goals and influences explain cycles of growth and decline. By introducing purposeful changes or influences into organizations, patterns of effects and effectiveness become observable. Katz and Kahn (1966) posited that energy must be continually imported into the system in order to maintain a pattern (p. 482). The "two basic criteria for identifying social systems and determining their functions are (1) tracing the pattern of energy exchange or activity of people as it results in some output and (2) ascertaining how the output is translated into energy which reactivates the pattern" (p. 482).

One may try to induce change by inputting one or more of the energies, but this does not always work in the way expected because not all energies are acceptable to the system at any one moment. Energy may be absorbed, rejected, or transformed; some kinetic energy may be stored, and too much potential energy may bring about a climatic release. The functions of the organization determine what happens to any energies input. Importations change the interval at which the organization's energy becomes adequate to induce change. Until there is sufficient energy to overcome resistance, wu-wei, quasi-acting asserts itself. Making a purposeful change requires knowing both the type of energy to apply and the states of all the energy capacities.

One's ability to make purposeful changes and anticipate effects is imperfect due to variability. Not only is the energy input variable, but when energy is released, the release may appear at unrecognized stress points in the pattern. Expectations are that patterns will be homogeneous, but when the pattern emerges it might not appear exactly uniform because of variability. Tangential shifts occur if there is sudden excess of either kinetic or potential energy.

Time plays a role in the appearance and form of patterns. The pattern may reappear unpredictably because exchanges are energized at different rates. Closure of a cycle may not reappear at regular intervals even when similar cycles are repeated regularly. Patterns will not always be recognizable in the short periods, but may take longer than expected to reveal themselves. Tight cycles can occur when an organization imports more energy than it expends and creates a habit. The cycle time gets shorter as the process is refined. Only when leadership recognizes the pointlessness of the repeated transactions will the habit be broken.

Wrongly suppressing variability, groups alter leadership decision strategies as situations dictate tapping one person's skills or another person's experiences. The organization needs to transform in order to restore resilience rather than repeat seemingly successful cycles where potential energy builds until external factors intervene.

REBALANCING AND THE TAO MODEL

In the adaptive enneagram pattern each of the two types of energies, kinetic and potential, are balanced. Information and Relationships are balanced by a judicious acting and nonacting in reflection. Rhodes (2010) offered using "wing points" (p. 99-101), the adjacent points, as a possible adaptive mechanism to maintain resilience. These wing points "give us additional resources from which to draw" when the energy "isn't sufficient for some reason" (p. 101). While wing points remain controversial relative to the use of the enneagram in explaining personality behavior, the variations that exist in organizations caused by the appearance of dominant energies becomes almost a necessity when examining organization process modifications.

We think the Tao model provides a different path to resilience. Energy flow passes continually through the Tao providing an opportunity for reflection. It is our thinking that, during reflection, what the Buddhists call absolute (concrete reality) knowledge is balanced with what the Taoist philosophers call relational (abstract) knowledge. Both knowledge forms shape the way the organization embraces, and is embraced by, internal and external relationships. But, if the knowledge becomes polarized toward one knowledge form, the energy flow will fall out of balance affecting the relationships.

In the out of balance examples presented we see the leaders' attempts to rebalance. Different leadership decision making employs different forms of energy. In the Strategic Leadership example, there is a failure of the transactional leadership strategy. In the Command and Control Paradigm there is a failure of the managerial leadership strategy. These leadership strategies are heavily influenced by kinetic energy in the former instances and potential energy in the latter.

Transactional leadership is kinetic or aggressive. It is based on authority, responsibility, and accountability. Conduct is assessed with ethical measurements—trust, respect, "integrity, promise keeping, trustworthiness, reciprocity" (Burns, 2003: 28). Transactional leadership involves constant action; giving and taking, bargaining, deal making, and contracting. Burns (2003) described transactional leadership saying

It [transactional leadership] was the basic, daily stuff of politics, the pursuit of change in measured and often reluctant doses. The transactional leader functioned as a broker and, especially when the stakes were low, his role could be relatively minor, even automatic. . . . To change is to substitute one thing for another, to give and take, to exchange places, to pass from one place to another. These are the kinds of changes I attribute to transactional leadership. (Burns, 2003: 24-25)

Managerial leadership is potential in its energy form. Mary Follett (1926: 152-154) launched the concept of managerial leadership when she observed that issuing orders was not always responded to with obedience. The leader gives an order then waits. Follett believed that scientific management profited leadership by depersonalizing orders and discovering what order was most integral to the situation. If the order is not followed, new incentives are employed. Evolving systems theories hold that often "managers must simultaneously attend to integration and differentiation," and "focus on the task and interpersonal aspects" (Dennison et al., 1995: 525); what Knowles (2002) called, Work [5] and Structure and Context [8].

Rebalancing requires a different strategy for leadership that has both potential and kinetic properties—transformational. Individuals drive orderliness through both process and social relationships. Processes become ordered to achieve purposes in some sort of hierarchy (Simon, 1946: 117). Shared values, beliefs, and opinions become interrelated; forming organizational culture. Imbalance is corrected by formal measures (p. 132). Patterns of social interaction, causes and consequences, accumulate to become a process ethos redefining the organization. That new ethos alters the way the organization shares information and develops opinion and values. The organization adapts to the new reestablished social equilibrium.

Burns (1978) introduced the concept of transformational leadership. Tichy and Divanna (1986) called it "a discipline with a set of predictable steps" and "acted on within a framework" of three parts; revitalization, creating a new vision, and institutionalizing change (p. viii-ix). There are steps: trigger events, a felt need for change, a visioning activity, a mobilization of commitment, and institutionalized change (p. 29-31). In each step, an individual experiences endings, neutral zones, and new beginnings (p. 32-33). Resistance to change is futile, willingness to adapt is hopeful. The individual is required to disengage, "disidentify" (p. 64). People learn to overcome fear of the unknown and loss of organizational predictability (p. 75). The organization develops a new order based on the health of the technical, political, and cultural aspects of the system (p. 94).

In developing the Tao Model, a transformational tool was the goal. Figure 8 shows the Tao Transformation Model that we named the Loop of Virtuous Leadership (Fu, 2008: 38). The Loop suggests that leaders can engage people in rebalancing at any stage in the cycle. Burns (2003) stated, "I believe leadership is not a descriptive term but a prescriptive one, embracing a moral" dimension "is a moral necessity" (p. 2). In this, Burns declared the leadership in transformations are due to a combination of many working toward common goals (p. 71). Organizational structures Burns saw as collections of people organized in multiple "malleable, susceptible" systems (p. 216).

In Figure 8, the numbers now represent the transformation energy-flow sequence. While this figure begins from the Founding position, it could start anywhere in the cycle to reflect if one has achieved each required state before moving on to the next. This energy-flow moves from Founding to Sustaining, Innovating, Responding, Anchoring, Influencing, Strategizing, then to Implementing; and the loop returns to Founding to restart next cycle. The essential differences from the enneagram process are these. First, all eight energies are attended to prescriptively in the Tao. Second, the only background activity is the constant rebalancing in the Tao. Tao balances each energy that never reaches stasis; potential energies are accompanied by aggressive leader attention, and kinetic energies are allowed to play out without direct leadership

7. River - Strategizing - Learn
Pay attention to what happens and adapt

1. **Heaven - Founding - Intention**
   Protect and establish vision

5. **Mountain - Anchoring - Relationships**
   Align values, beliefs, and vision

3. Mist - Innovating - Work
   Attend to external drivers

Thunder - Responding - Information
Behave congruent with purpose

2. Earth - Sustaining - Issues
   Use & expand resources wisely

6. Wind - Influencing - Principles
   Promote new wants and needs

8. **Fire - Implementing - Context**
   Respond to outcomes and issues

Figure 8 *Loop of Virtuous Leadership (Fu, 2008: 38)*

intervention. Third, the sequence of the flow is altered by restoring Knowles's (2002) Information [6] and Relationships [3] as steps in the process as they were originally offered in the Living System.

Northouse (2001) described transformational leadership as "the process whereby an individual engages with others" creates connections, "raises the level of motivation and morality" and "tries to help others reach their fullest potential" (p. 132). Transformational leaders "have a strong set of internal values and ideals" (p. 136). Such leaders build on an ethical relationship of trust (p. 144) to instill others with a new sense of identity (p. 145). The leader satisfies both his/her and followers' needs (p. 146). Northouse indicated that transformational leadership is often interpreted as an "either-or" approach instead of a matter of degrees by aspiring leaders and often elitist and undemocratic in its implementation (p. 147). A transformational example of the Loop of Virtuous Leadership might be useful.

We begin this example with Founding. Founding energy relies upon the leader having a vision of what needs to be done, but also requires that the vision be visible. If it is an existing vision, that vision must be diligently restored; with time all visions become tarnished. Visions that are espoused and not shared by the collective will not transform the organization. The energy is kinetic as leaders must establish a foundation for transformation to assure there is movement toward the inherent organization goal(s).

Sustaining (potential energy) the restoration or inauguration of the vision makes resource issues apparent. This is not the sole responsibility of the leader, but of the organization. Leaders refrain from controlling, but Sustaining and developing. Resources must be employed consistent with the vision to Nurture the goal(s) and develop the organizational capacity to sustain transformation. The leader guides (potential energy) how Resources are distributed.

Innovating builds potential energy. External factors will invoke creativity perhaps to turn the organization toward greater efficiencies, leverage available talents, or acquire technologies it can afford. Potential energy is increased when, for example, decisions are made to replace outdated constraining resources and add capacity for new activities.

Responding is kinetic activity. Leaders engage organizations, actively listen and respond to what those voices are saying, allowing the message to become meaningful to all. By attentively Responding (kinetic energy) to feedback, either consent or dissent, all who engage will become champions and committed to the transformation. Maslow (1943) described the process of what happens as needs and wants are fulfilled.

Anchoring links people to values (kinetic energy). If anchoring does not take place, the transformation will flounder. Energy is expended in changing the processes, policies, procedures, and even the organization must align to the values and organization purpose. Traditions that are not useful that hide faults or errors are eliminated. If organization defensive routines (Argyris, 1990) are employed they are aggressively dismantled.

Influencing (potential energy) does not coerce rather it allows new wants to surface and needs adapt. Both internal and external stakeholders will adjust anticipating things that will have to be addressed during the transformation. Obligations and commitments need to be fulfilled. Wise leadership anticipates that new needs will form even as the old ones are satisfied.

Strategizing reshapes direction (kinetic energy). Every lesson from past changes needs to be captured and communicated in plans. Human resources are allocated and trained. When necessary the organization is restructured to fit the transformation purpose and people may be reassigned to appropriate positions. Budgets are drawn and contingency plans made.

Implementation sounds like a kinetic activity, but for the leadership it mostly passive (potential energy). Challenges will arise and they will be attended to, often by implementing tactical contingency plans, and guidance. The organization will need illumination to guide its path from time-to-time. Allowing challenges to drift only implies a lack of concern, so oversight (potential energy) is constant. Good behavior is acknowledged and rewarded.

Just as in the enneagram, the flow of energy returns to the beginning to start a new transformational cycle. The world, however, does not tolerate quick sporadic cycles.

SUMMING UP

Capra (1975/1991) explained that Tao, has a natural order to its flow; as does the Process Enneagram. Change is not a consequence of an incidence of applied energy but rather a tendency innate to all things and situations. "The movements of the Tao are not forced upon it, but occur spontaneously" and responsively (p. 116). Peoples' actions then need to synchronize with the flow of Nature's energies at that point in the flow. Where the flow is focused the energy should be responsive. If the energy called for is action then the energy to be applied is kinetic. If the energy required is allowing, waiting, or passively contemplating, then potential energy is called for. Restoring balance requires trust in one's true native and intuitive intelligence (p. 116-117).

The Tao Model and the Process Enneagram hold a great deal of promise as they accommodate the complex environments that face leaders today. We believe that the contributions made by them will enhance understanding of why there is not always an easy fit with reality and why different prescriptive actions and leadership strategies do not always yield perfect solutions. We encourage further use of both and hold hope that future applications with an understanding of how the energy works will prove beneficial.

REFERENCES

Bohm, D. (1980). Wholeness and the Implicate Order, ISBN 07448000085.

Burns, J.M. (1978). Leadership, ISBN 0060105887.

Burns, J.M. (2003). Transforming Leadership: A New Pursuit of Happiness, ISBN 0871138662.

Capra, F. (1975/1991). The Tao of Physics: An Exploration of the Parallels Between Modern Physics and Eastern Mysticism, ISBN-877735948.

Denison, D.R., Hoojberg, R., and Quinn, R.E. (1995). "Paradox and performance: Toward a theory of behavioral complexity in managerial leadership," Organizational Science, 6(5): 524-540.

Follett, M.P. (1926). "The giving of orders," in J.M. Shafritz, J.S. Ott, & Y.S. Jang (eds.), *Classics of Organization Theory*, ISBN 0534631568, pp. 152-157.

Fu, C. (2008). Energy-Flow—A New Perspective on James MacGregor Burns' Transforming Leadership: A New Pursuit of Happiness, doctoral dissertation, Antioch University, Yellow Springs. Ohiolink ETD: http://rave.ohiolink.edu/etdc/view?acc_num=antioch1218205866.

Katz, D. and Kahn, R. L. (1966). "The social psychology of organizations," in J. M.Shafritz, J.S. Ott, and Y. S. Jang (eds.), *Classics of Organization Theory*, ISBN 0534631568, pp. 480-490.

Knowles, R.N. (2002). The Leadership Dance: Pathways to Extraordinary Organizational Effectiveness, ISBN 0972120408.

Lao-Tzu. (1891). The book of Tao (Tao Te Ching) (J. Legge, Trans.). (Original work published 500 BCE) Retrieved from www.sacred-texts.com/tao/ taote.htm.

Northouse, P. G. (2001). Leadership: Theory and Practice, ISBN 0761919252.

Rhodes, S. (2010). Archetypes of the Enneagram: Exploring the Life Themes of the 27 Subtypes from the Perspective of Soul, ISBN 9780982479216.

Simon, H.A. (1946, Winter). "The proverbs of administration," in J.M. Shafritz, J.S. Ott, & Y.S. Jang (eds.), *Classics of Organization Theory*, ISBN 0534631568, pp. 112-124.

Siu, R.H.G. (1978). "Management and the art of Chinese baseball," *Sloan Management Review*, ISSN 0019-848X, 19(3): 83-89.

Vaill, P.B. (1989). Managing as a Performing Art, New Ideas for a World of Chaotic Change, ISBN 1555429638.

Wilhelm, R. and Baynes, C.F. (1997). *The I Ching or Book of Changes*, ISBN 069109750X.

Chapter 9

PROCESS MAGIC WITH THE ENNEAGRAM

Steffan Soule

This article instructs the reader to follow a card magic routine that demonstrates how the enneagram can be used to understand, track, observe and improve a whole system. Using a deck of cards, the reader will experience a real life example to see the qualities and connections made by the enneagram. Since these fundamental principles exist in every completing process, the reader will be able to use this routine later as a way to recollect the whole systems approach of the enneagram.

INTRODUCTION

A well-run process feels like magic. When your plan comes together and heads toward completion, when insights, new actions and refinements enter your process, you become the magician.

In fact, every magician learns to follow the path of a well-run process. The sequence of events during a magic effect preserves a secret and creates a transformation in the minds of the viewer. If the audience sees an incongruity, they will not be amazed; therefore efficiency must predominate when a magician guides amazement. Learning the sequence of a magic effect helps us understand efficiency and transformation.

Magic is not about the speeds of the hand and the eye; it's about delivering an efficient process. To amaze your friends with a top notch, audience tested magic trick, all you have to do is follow a perfect pattern. Leading spectators down the garden path to the beauty of wonder requires understanding the whole pattern, seeing what will remain hidden, and guiding specific parts to come together.

This sounds like a job for the enneagram. As you know, the enneagram depicts many forces combining into a transformation factory. For those versed in the language of the enneagram, what I call the Nine Term Symbol (NTS, Soule, 2011), or in the language of the Process Enneagram™ (PE, see Knowles, 2002), we can outline a magic effect to see efficiency in action. Even as we examine one simple magic trick, the entire NTS or PE can be seen running the show. And then, when you perform the effect or even rehearse it, you can experience the magic of the enneagram as a tool for understanding process. This will show you the power of using this as a tool to observe and keep track of a process.

As we proceed, you will learn a closely guarded principle used by magicians to amaze people. You will also learn a process that can be outlined and reviewed on two enneagrams, the NTS and the PE mentioned above.

The process requires that you have a deck of cards, you know the secret, and you remember the sequence. Because it follows the whole-system-structure of the enneagram you are learning a pocket size example of process transformation. You can use this to quickly review and add to your understanding simply by carrying a deck of cards and presenting this routine with creative attention. That is the biggest secret of all, creative attention. Anyone who has come to point-seven either within the Nine Term Symbol or within the Process Enneagram can see that creative attention (at the point of Deep Learning or along the line of Restoration/Re-evaluation) leads to insights and the ability to integrate improvements. We will see more about this when we come to the performance process.

As for the magic trick, please keep it a secret. A magician never reveals the secret of a trick. You may study the secret with others with the intent to perform, but you may not reveal the secret in casual conversation, merely to satisfy curiosity, or for profit. The secrets of the enneagram however are meant to be shared, so if you find the following description to be useful in learning those secrets, by all means share it.

For your reading and process transformation pleasure, you will need to have your deck of cards near by... and an audience. Yes, you can do this simple magic effect with business cards, but it may lose its impact. As for your audience, we will begin with your imagination.

Picturing The Scenario

Imagine a restaurant filled with twelve of your friends or coworkers, but they are spread out at tables of four. They are in different areas of the restaurant so you can repeat your performance at each of their three tables without anyone overhearing. Or for another challenge, imagine you are sitting in a restaurant. Look over at tables of strangers and picture what it would be like to walk over and perform this right at their table. It is understood that the restaurant is supportive and that you are authorized to perform. Then you go to another table, and then one more.

Are your cards wet? If this question seems to come from nowhere, we should look at one of the major obstacles to card magic. There is often water on tables. It usually comes from cold water glasses and sometimes little spills. If your cards get wet at all, they will stop working, and your show is over. If you dry the table with a cloth, it looks odd. If you place a card mat onto the table, it helps, but it adds another prop to your prop bag. Now you are carrying more than a deck of cards. You can look for the dry area and only go there. The plot thickens. Maybe the magic is in the details.

Before we go any further, perhaps we should look into more of the difficulties every magician encounters during a performance, even one as simple as the trick you are about to master.

People are not easily amazed. We live amidst powerful technologies that should astound us, yet we take them for granted. We expect to be entertained even when we watch the news, and we imagine that for every question there's an immediate answer awaiting us online. Where's the magic these days?

I sometimes perform magic in an elegant setting at an historic hotel. I roam through the lobby presenting table-side magic. Magicians call it table-hopping. Some come specifically for the magic and some don't even know it's magic night. When I approach a table of unsuspecting guests, I do not know for sure if they will want to watch. I must discover if they have an interest, an opening. I am armed with some of the best magic in the world, but they don't know that, and they're not expecting a show at their table. They might consider it an interruption. The reputation of the hotel, my reputation, or just the expectation of getting an entertaining experience might help me break the ice.

When I approach them and begin to present myself, I don't say, "Want to see a trick?" I say something more appealing like, "We are aiming for the impossible tonight, and I am going to see if it happens with you, here, right now." As I open my mouth, I am sensing their reaction; a tension between us is launched, and there is potential for a relationship. I am about to deliver a powerful experience or be dismissed, inter-rupted, or upstaged by any number of conflicting values. If it were me sitting there, I would clear the entire table to make room for the guy and his props, turn off my cell phone, and tell the waitress to hold everything, even if it's hot. But most people do not value live entertainment with the same zeal as me, especially when it's free and just happens to arrive at their table.

The hotel wants to offer it, I want to give it, and for some, the experience will become a life long memory. But I don't know when that kind of magic will happen. I must be prepared to wind up my intro, do something quick and move on, or to set the table for the best show in town. You'll see for yourself if you learn the magic effect I describe here and then dare to perform it somewhere. That's a setting filled with risk.

Let's say I am met with acceptance. It usually happens that way. I begin the effect; my real job is now underway. I can't just act out a memorized script and expect the job to work itself. We are now in a creative interaction where rehearsed scripts and spontaneous actions come together in new ways. The structure of the event emerges unpredictably. One of the spectators says something funny, or the waiter delivers a plate or a drink right at the moment when I am requiring their full attention. The facts of the moment matter. Every time they look away, they might miss something crucial, so sometimes I have to pause and wait for them to look back without seeming like I am waiting. There is a flow of information back and forth between us. Did I touch the envelope after I placed it on the table, or was it completely out of my hands the entire time? When their signed card appears inside that envelope, was it magic, or did I switch it when the waiter came? Will we all agree to keep our attention on my hands and my props, or will we look away at the waitress or attend to the text message streaming on our little screens? If I verbally assert my needs—which are merely those of the process—will I come off like an artist in need of attention, or will it be funny because I appear to make fun of the distractions? I have learned to incorporate distractions into the show, but often it does not stop my spectators from taking their eyes off my visual art for way too long to maintain the intended experience.

Once the first magic effect has been presented, was it good, did we like our experience together? They can indicate when they've had enough in many ways, and I can easily move on since there are other tables. But usually we are just getting started, and now we have a new identity. Something amazing just happened. We all know it, and we all want more. We are entering into full show mode, and my new friends are about to see the best magic they could hope for. This is when you will want to have a few more powerful effects up your sleeve, but let's begin with one.

Steffan Soule's Even And Odd Miracle

Try this next effect and you will immediately wish to share it with others. But, before you do, practice, practice, practice.

Remove 8 red cards and 8 black cards from a deck of cards. They can be face cards or spot cards, it does not make a difference.

This is a mysterious ritual using cards. Shuffle or mix the eight red cards and place them in a face down pile on your left. Shuffle or mix the eight black cards and place them in a face down pile on your right.

Each pile has an even number of cards. This trick is called Even and Odd for a reason: one of the piles is about to become odd.

Which one of these piles that becomes the pile with the odd number of cards is ultimately up to you, but for this demonstration, let's say it is the black pile on your right. To make it the odd pile, you must remove one card, so take the top card and place it back in the deck. The pile on the right which is comprised of black cards is now the "odd pile". It actually has an odd number of cards.

Now take the red cards (the even pile on your left) and deal these cards into two even piles starting on your left, then to your right, back and forth, until all of the cards are dealt. Say "Even" as every card is dealt.

Yes, every time you deal one card, you deal another to the other pile. As you deal, you even out each pile, so you must actually say "Even" out loud to reinforce this fact.

Next take the black cards (the odd pile) and deal them left to right on top of the two piles, back and forth, saying "Even", until there is one card left. This is the odd card. Again you say "Even" as every card is dealt except for the last one. Do not say "Even" as you deal the last card. This one time you say "Odd", and you place it on top of the right hand pile. This is, after all, the side we chose for the odd pile.

For the even and odd ritual to be complete, you must cover the odd pile on the right with your right hand, pause while you say "Odd" and then move your right hand to the left hand pile and say "Go". You should repeat this again but this time say, "Odd Card Go Here" while you point at the odd pile on your right and then to the left hand pile. While you say this and point, you are willing the odd card to travel mysteriously to the left hand pile of even cards.

The magic is done! Take a pair of two cards from the right hand pile and say "Even". Put them in your pocket. Take the next pair (again saying "Even"), and put them in your pocket. Saying "Even" with each pair, take the next pair, and so on, until you see for sure that the odd card has vanished.

Now take a pair from the left hand pile and put it into your pocket saying, "Even". And, continue repeating this with pairs of cards until you find the odd card that has invisibly traveled.

Can this ritual be explained or is it beyond understanding? Only the ultra curious will resolve this apparent impossibility. I will leave it at that, but, I want to assure you, this effect can be used to entertain and mystify even the most sophisticated audience, so, please guard it well and use it only if you have practiced the method and your presentation numerous times.

Your Script

The script you use when you present this effect must be precise. You must hide the idea and the fact that the "odd pile" at the end actually is comprised of an even number of cards. You must never count the cards by number except at the beginning when we need to get eight blacks and reds. The idea of making one pile odd and then reinforcing this image by calling out "Even" on every deal (except for the last one) fools even learned scientists.

But let's not forget, people want to be amazed rather than fooled. By referring to this as a "magical card ritual," you take the heat off of the idea that you are an accomplished magician or that you are going to perform something over the top. It is best to downplay expectations, and one way to do this is to say that you are not going to do the magic, but they are! All of the magic will be done in their hands. You will not even touch the cards.

The script to introduce the ritual might sound like this:

"We are now going to attempt an ancient ritual involving unseen forces. If this is successful, you yourselves will cause a playing card to travel invisibly from one place to another."

You can perform this effect for many people at once as long as you can get them to follow your verbal instructions!

"Please gather 8 red cards and 8 black cards from a deck of playing cards. Remove them from the deck and place the deck aside. Shuffle your reds and place them on your left. The heart side"

Notice how the number 8 is not repeated when giving the shuffle command. Shuffling here does nothing but it creates a psychological sense of randomness. It gives them a feeling that there is no way you could control anything now because there is chaos at play! Mentioning they are on the left and that this is the heart side is a psychological suggestion to forget about numbers and enter into mystery. The global financial industry has over used this technique, but we are engaging in entertainment and creating a positive experience, so a little suggestion is fair game. It is artistic license implicit in the relationship of magician and audience. Unlike most professions, magicians agree never to hide the fact that what we do is a show. In most parts of the world, you do not have to admit you are going to trick people; as long as they know you are a performer, a magician or even something unheard of like a "ritual guide," it is understood.

"Shuffle the black cards. Mix them into random chaos; shuffle, shuffle, shuffle. Okay, now set them down on your right. One of these piles will become odd. You choose! Point to the pile that you want to become ODD."

Here you must learn how to perform this effect with either choice, left or right. It will take extra rehearsal to make this trick "ambidextrous," but it's a good exercise. The easiest way—which also works if you are performing for many people at once—is to have them deal the EVEN pile out left to right, deal the ODD pile out the same way and place the ODD CARD onto the side that they previously picked to be ODD. The mechanics of this will sink in the more you grasp the simple principle behind this effect.

"Great. To make this your ODD pile, you just remove one of the cards from your odd pile and place it aside onto your deck of cards. You now have an EVEN pile and an ODD pile.

"Now take the EVEN pile and deal these cards into two piles starting on your left, then to your right, back and forth, until all of the cards are dealt. Say 'Even' as every card is dealt.

"Yes, every time you deal one card, you deal another to the other pile. As you deal, you even out each pile, so you must actually say "Even" out loud to the whole wide world, to the universe.

"Next take the ODD pile and deal them left to right on top of the two piles, back and forth, saying "Even", until there is one card left. This is the ODD CARD. Do not say 'Even' as you deal the last card. This one time you must say 'Odd', and you place it on top of your ODD pile. This is, after all, the side you chose for the odd pile.

"For the even and odd ritual to be complete, you must cover your odd pile with your hand, pause while you say 'Odd,' and then move your hand to your even pile and say 'Go'. Now point to your even pile and say, 'Odd Card Go Here'. You have to will the odd card to travel mysteriously to your pile of even cards.

"The ritual is nearly complete! Take a pair of two cards from your original ODD pile and say 'Even'. Put them in your pocket or onto your deck. Take the next pair (again saying 'Even'), and put them in your pocket or deck. Saying 'Even' with each pair, take the next pair, and so on, until you see for sure that the odd card has vanished.

"Now take a pair from your original EVEN pile and put it into your pocket or onto your deck saying, 'Even'. And, continue repeating this with pairs of cards until you find the odd card that has invisibly traveled. You have done the magic!"

The above script will work for most occasions but you must be very open and allow for creativity to enter the scenario. After all, when presenting something as intimate as a ritual, you would expect to have creative interactions with your participants. If you are with friends, you will be adding a new experience to your memories, and if you are strangers, you might be making new friends. But in every case, you will have to remain attentive and creative as the process unfolds and the moment of magic takes hold.

Pondering The Symbols

How does all of this look when viewed using the Process Enneagram and on the Nine Term Symbol? If you know how to read these symbols, this will quickly make sense, and if you are new to these ideas, perhaps this will validate that the time you put into exploring the books explaining each of these methods is worth your effort. Consider this: if these symbols really do relate to a process as simple as this little magic trick, then maybe they can scale to any process. They can and they do! That is the magic of the enneagram symbol. So let's look first at the PE and then conclude with the NTS.

Magic Effect On The Process Enneagram (PE)

The Process Enneagram shows how we must move from our Intention to our Values in order to help us work with Tension Issues (see Figure 1).

The kinds of Tensions that come up during the first part of a magic effect are numerous. I have described some above. To relate to them, you can picture what it is like to approach someone with the intent to entertain them with a card trick. Now you

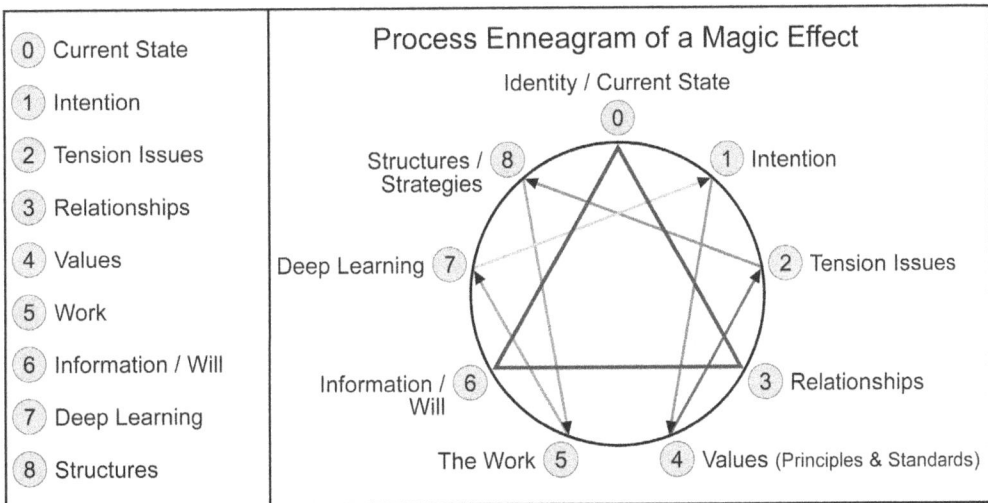

Figure 1 *Process Enneagram List and Symbol*

may not call it a card trick because that would diminish the mystery, but that is what it is nonetheless. What if they would prefer to keep talking privately with their friend? What if they act like the trick is obvious and do not appear to be amazed? What if they tell you they have already seen that trick? The list goes on and on, but there may be some common values. Maybe they want to be entertained. Perhaps they came to see the magic or they became interested when they heard the applause from another table across the room. Maybe they are fascinated just to see a live magician. There will be something in common when your circles intersect.

By looking from the Tension Issues over to the Structure, you are visualizing what general form and sequence will best suit this specific person or group in this moment. It may be exactly as rehearsed or it may need adjustment according to the many factors contributing to the Intention, the Identity and all possible Structures.

When the trick is underway at point 5, the degree of attention the audience puts onto or into the process will ultimately serve to enhance the outcome. This is point-six. The term Information refers to how easily all the information of the trick is as-similated into the event. The setting, the mental, physical and emotional states of the audience, get processed as Information and also as Will. The Will is, in the case of the magic effect, the attention of the audience and also the magician's power of attention and perseverance during the performance.

As the effect is finished, the process continues as the audience has a reaction and the magician looks into the result to see how it compares to the original Intention. This is point-seven when Deep Learning begins. Sometimes the audience will feed a funny line into the sequence that can be re-written and used for every performance. This requires listening during The Work (point-five) when the routine is underway; listening for special moments so they can be recalled at point-seven.

Once the effect is experienced the Current State has changed and the group has a new Identity. Figure 2 shows the Process Enneagram of this magic effect. For a full un-derstanding of the magic effect and how the enneagram of a magic effect completely coincides with reality, one may expect to actually work at the effect described, perform it and then re-read this article, possibly a few times. Words fall short. This is a hands-on, ex-periential approach to learning. The words we are using help, and the actual experience combined with an effort of pondering the symbols here expands ones understanding.

The central idea of the Process Enneagram was developed about 20 years ago by Dr. Richard N. Knowles, and it represents a breakthrough in applying this tool. His

Magic Effect (PE)

Identity / Current State

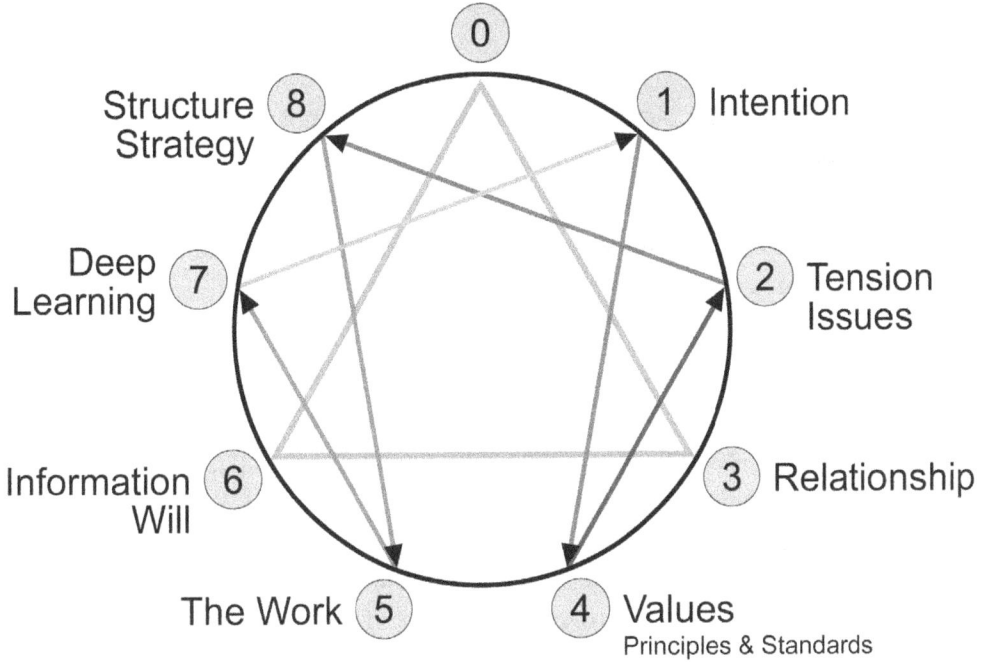

Figure 2 *Process Enneagram*

insightful book, *The Leadership Dance* is available from Amazon, and is quite simply a must read for enneagram practitioners.

We are about to look at the Nine Term Symbol (NTS). In my book *Accomplish The Impossible* I call it the NTS to distinguish it from ways of using the symbol that have nothing to do with process work. Here we could simply call it the enneagram, but the system developed with Richard N. Knowles—which I have utilized above—is slightly different, so we will continue with the name "NTS" or Nine Term Symbol.

The Nine Term Symbol differs from the Process Enneagram in terms of its emphasis on the Compass. The Compass is the six-pointed figure which demonstrates the qualities needed at every step of a process. Along with these specific qualities, the Compass simultaneously shows which points of a process need to be connected. These qualities and connections deepen with attention and repetition. While the Process Enneagram also relies on the connections made by the Compass, the NTS defines and emphasizes what qualities these connections contain in a process that reaches completion. Also, the NTS shows many different versions of the Triangle for different types of processes whereas the Process Enneagram guides us to a specific transformation engine that uses a specific Triad embodied in a universal pattern. Each of these two methods demonstrates an understanding of the laws symbolized by the emblem which means that the Process Enneagram and the Nine Term Symbol are always, in essence, compatible.

Incidentally, my book, *Accomplish The Impossible*, is not about magic tricks. It is about the enneagram and understanding whole systems and process transformation, and there are hundreds of examples in my book pertaining to business, art, science and life. The book simplifies the enneagram as a tool for understanding process.

Magic Effect On The Nine Term Symbol (NTS)

Now let's return to the study of the Symbol through the practice of our little magic routine; this time with a focus on the NTS and its magical Compass. A study of Figure 3 will help us identify the qualities embodied in the Compass and the pathway of these qualities through the process.

Many business executives who watch me perform up close at a social hour will make jokes like, "Hold on to your wallet Sam" or "Check your watch," as if I am a pick pocket! A magician is different. In fact, magicians are more honest that some business experts because by definition, a magician admits up front to using secrets. The Secret (point-one) in a magic effect does not equate to deception in business, no! Every process has possibilities which must be emphasized for success and possibilities which must be resisted so they do not ruin the effect of the process. In business you have the group that is coming together for an intention, and some needs of the group

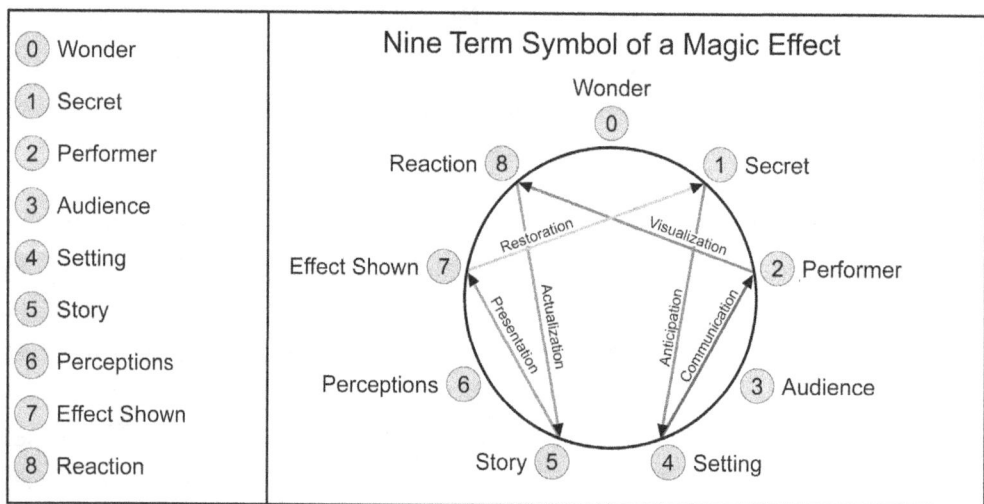

Figure 3 *NTS List and Symbol*

must be kept out of the present moment for the intention to be clear and move forward. The needs are still there. Later the group might have to respond to those needs, but not in this go around if the process is to become highly successful.

The Secret must be presented properly. If you are the magician, you must (following the pattern of the Compass) Anticipate the Setting and Communicate the needs of the Setting to your role at point-two. At point-two, you are ready to go as the Performer. Your effects are ready (in your case maybe only one effect is ready), and you enter the scene. By Visualizing the best possible Reaction from the transformation that will occur in the Perceptions of your Audience, you will be able to strategize and Actualize the best Story at point-five. This may sound esoteric, but it is super practical. This is how the best shows are achieved. The aim has to be kept in mind. If I loose sight of creating the best possible Reaction and allow other concerns (like what I am going to do tomorrow) to take over, the effect can suffer. By keeping my attention on the final aim, I will use all possible means available at the Setting and all ways of adapting my Story to blend with the needs of the Audience and the particular types of Perceptions they are revealing to me throughout the interaction.

When someone says, "Can I shuffle right now, or would that mess you up?" I might reply, "Not a problem; that would be a whole different show, which we can do right after this. Let's stay focused on exactly what I am saying here, so we can see if the magic will happen, then we can do one that you invent." Or I might say, "Yes, you got it; that would really mess me up!" What I say, how I respond and what posture I take depends on their attitude! There is a feeling level to every moment in a magic show.

Once the Effect is Shown (the trick is performed), I am already restoring the cards and getting my materials back in order so I can enter into the next effect or hide the Secret or begin the next guided moment. Everything can be intentional when you are the performer. The more intentional, the greater the magic!

I look toward their Reaction to see what level of Wonder has emerged and this informs my next choices. Do I continue? Which effect do they need next in order to bring them to the highest level of entertainment? (See Figure 4.)

As you work with this simple magic effect, I believe you will find all aspects of the Process Enneagram and all parts of the Nine Term Symbol to be useful in furthering your understanding of the whole process. You will see gaps in your Presentation (line 5-7) or in the Work (point-five) that must be filled the next time you approach the routine. You may find many examples of how the symbol helps you see what is needed and what is not needed within your process. Indeed, as a professional magician, I

Magic Effect (NTS)

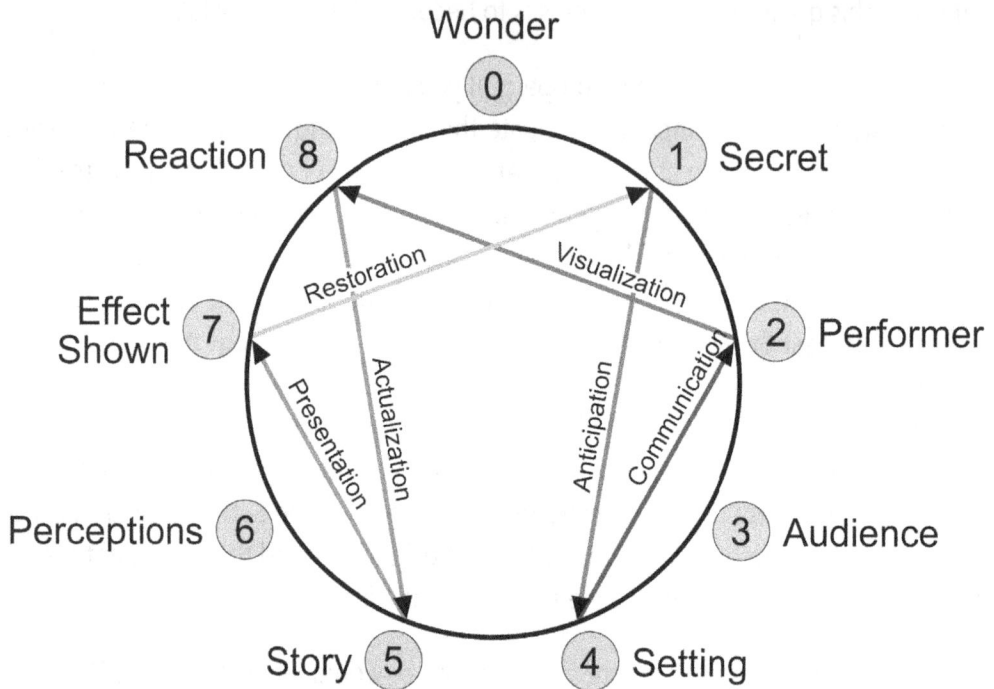

Figure 4 *Nine Term Symbol*

have found nothing more amazing than the enneagram for revealing all of the necessary components and linking them together into one concise whole. This can only be discovered through personal repeated efforts. It doesn't happen by magic.

REFERENCES

Soule, S. (2011). Accomplish The Impossible: The Six Secrets of Sustainability and Transformation for Business, ISBN 9780984240517.

Knowles, R.N.(2002). The Leadership Dance: Pathways to Extraordinary Organizational Effectiveness, ISBN 9780972120401.

www.ingramcontent.com/pod-product-compliance
Lightning Source LLC
Chambersburg PA
CBHW061817210326
41599CB00034B/7023